CONCEPTS AND TECHNIQUES IN MODERN GEOGRAPHY No. 33

AN INTRODUCTION TO DIMENSIONAL ANALYSIS FOR GEOGRAPHERS

by

Robin Haynes

CONTENTS

ACKNOWLEDGEMENTS

I am grateful to Vince Gardiner and an anonymous reviewer for their helpful comments.

1. EQUATIONS AND DIMENSIONS

All geographers need to be able to understand equations. While it is true that a great deal of geography cannot be described in the language of mathematics without being made trivial, much remains that can. An equation, where it is appropriate, provides the most concise and informative summary possible. To understand equations you need to know the rules of mathematics. But that is not all. Unlike the sterile exercises in some mathematics books which concentrate on the manipulation of abstract symbols and numbers, scientific equations refer to the working of real events in the world outside. Skill in handling numbers and symbols will not necessarily tell you what the final product actually means. It is quite possible for an equation to be mathematically correct and yet be scientifically (or geographically) nonsense. Scientific equations are built not on abstractions but on measurements of actual phenomena. Making sense of scientific equations depends on appreciating the difference between a number and a measurement and an understanding of the implications of adding or multiplying measurements. Dimensional analysis is the study of measurement and its influence on scientific relationships. The techniques of dimensional analysis are not difficult to learn and yet they are extremely helpful. They have been used by physicists and engineers for many years to derive theoretical relationships, check experimental equations, interpret the results from scale models and convert between different systems of units. They are also applicable in geography. Once you know the rudiments of the technique, you might never look at an equation in the same way again!

1.1 A simple relationship

One of the attractions of dimensional analysis is that it enables you to predict how variables are related to each other. A few moments' thought might be enough to solve a problem that would otherwise take a week's fieldwork. To prove this claim, let us look at the relationship between channel frequency and drainage density.

Channel frequency and drainage density are both measures of the quantity of stream flow in an area of land. Channel frequency is the number of streams per unit area while drainage density is the total length of streams per unit area. If we were to measure these two properties in different places, we could plot a graph to show how one responds to changes in the other. The points on the graph would tend to fall in a line if channel frequency and drainage density are related; that is, if one can be predicted from the other. A large variety of different relationships might be possible: the line which summarises the trend of the points could be straight, but it could equally well be curved. Figure 1 shows some of the more common possibilities. According to which possibility your graph resembled most, you could write an equation for the relationship between channel frequency F and drainage density D:

(i) $F \propto D$ or $F = kD$

(ii) $F \propto \sqrt{D}$ or $F = k\sqrt{D}$

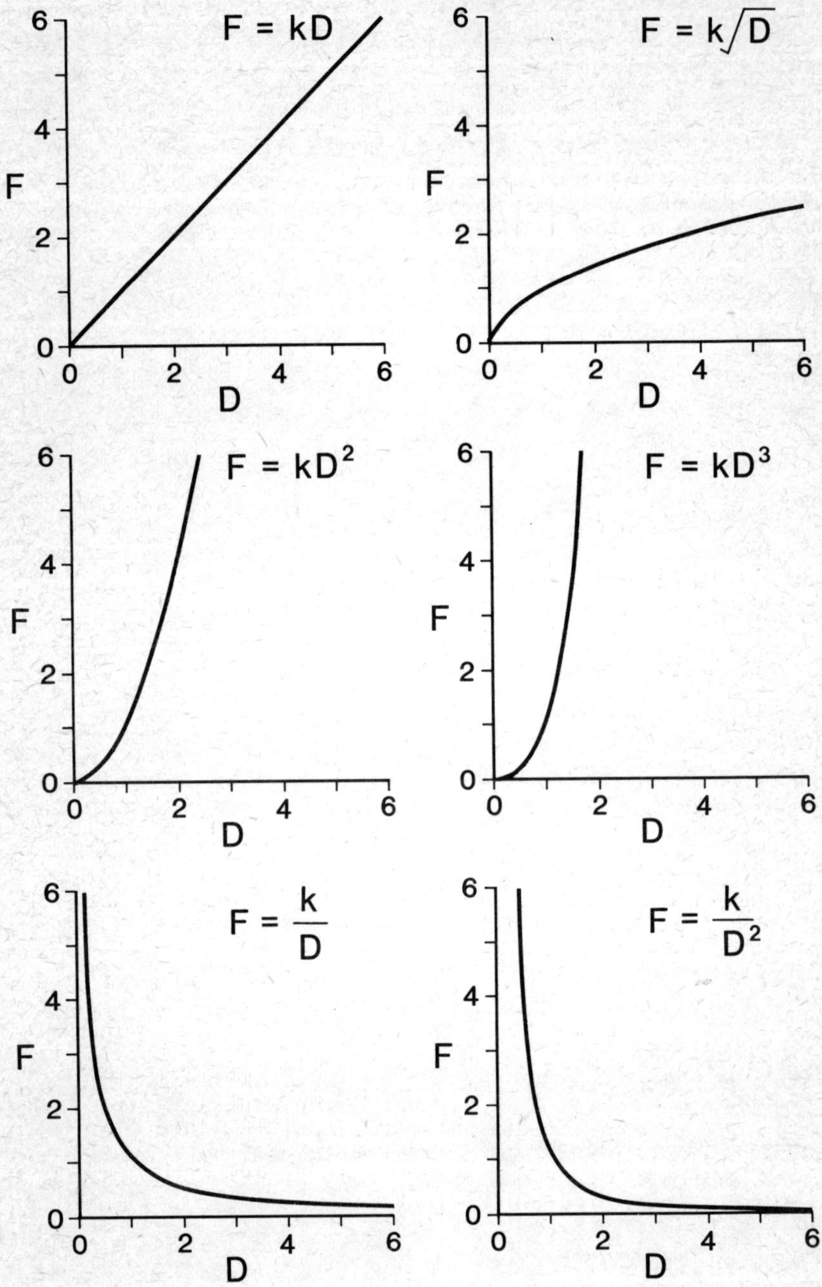

Figure 1. Various relationships

(iii) $F \propto D^2$ or $F = k\,D^2$

(iv) $F \propto D^3$ or $F = k\,D^3$

(v) $F \propto D^{-1}$ or $F = \frac{k}{D}$

(vi) $F \propto D^{-2}$ or $F = \frac{k}{D^2}$

The symbol $\propto$ means 'is proportional to'. If F is proportional to D (as in (i) above) then F responds in the same ratio to any change in D. Being proportional is not the same as being equal. To make the statements of proportionality equations then a constant of proportionality k must be added. The constant k is the number of streams which correspond with a drainage density of one unit. It is added to make sure that the numbers on both sides of the equation come out the same. In all the graphs in Figure 1, k = 1 for simplicity.

Figure 1 does not exhaust all the ways in which two variables might be related but it is probably baffling enough with just six possibilities. Which one of them is correct? Only when you know the answer are you able to start predicting one variable from values of the other. As we have seen, you could find out by measuring the channel frequencies and drainage density in a large number of places, and matching your results against Figure 1 to see which relationship fits best. This would take a good deal of effort. But why do unnecessary work? There is a short cut. Only one equation in the list above can possibly be true; none of the others have a hope of being correct. Dimensional analysis can quickly identify the true relationship (it is actually (iii)) without the need for you to do an experiment. To see how, we must go back to the basics of measurement.

1.2 Primary and secondary dimensions

All objects have properties that can be measured. A river can be described in terms of its length, the area of land it drains or the amount of water it discharges in a certain time period. A rock can be described by its weight or its density. The atmosphere has the properties of pressure, temperature and volume, which can be measured. A city can be measured in terms of its population, its age or its latitude and longitude. Any property that can be measured is a dimension.

A unit defines the magnitude of a dimension. The property or dimension of length, for example, may be measured in metres, kilometres, inches, miles and even nautical miles. Measurements consist of a number multiplied by a unit. The length of a river cannot be 8, but it can be 8 km. Varying the unit (from, say, kilometres to miles) affects the numerical value of the measurement (8 km is the same as 5 miles) but it does not affect the dimension which is being measured.

Some measurable properties can be regarded as basic or fundamental, to distinguish them from other properties that are made up of combinations of the basic properties. The property of length is usually thought to be fundamental because the properties of area and volume can be derived from it. An area, after all, is simply a length multiplied by another length, while a volume consists of three lengths multiplied together. Length is called a primary dimension, and area and volume are known as secondary dimensions. Two other primary dimensions are mass and time. Together, the three primary dimensions of mass, length and time can be used to define a high proportion of the properties which are of interest in physical science. The property of speed, for example, is a length dimension divided by a time dimension. Density is mass divided by volume (the product of three lengths), and so on.

To reduce the verbiage, let us introduce a shorthand system. We shall denote the mass dimension by [M], the length dimension by [L] and the time dimension by [T]. This gives us a concise way of referring to the other properties that are dependent on these three primary dimensions. Area can be written as [L x L], or $[L^2]$ for short. Volume is therefore $[L^3]$. Density (mass divided by volume) is $[M \div L^3]$ or $[M/L^3]$. This can also be written $[ML^{-3}]$. Speed is [L/T] or $[LT^{-1}]$, while acceleration (length per time per time) is $[L/T^2]$ or $[LT^{-2}]$. The property of force (mass multiplied by acceleration) becomes $[MLT^{-2}]$. Pressure (defined as force per area) is $[MLT^{-2}/L^2]$ which compresses to $[ML^{-1}T^{-2}]$.

By this time, you may feel you need a refresher course on exponents. If so, you should refer to the rules given in Table 1. Note that the square root of X can be written $X^{0.5}$. To divide by X is the same as to raise X to the power of -1. Dividing by X^2 is the same as multiplying by X^{-2}. When the same quantity appears more than once in an expression with different exponents, the exponents are easily combined or cancelled. The special case given at the end of the table is when a quantity is divided by an identical quantity. This cancels out the original quantities and always produces the answer 1. We shall be meeting examples of this shortly.

TABLE 1. The rules for exponents

$$X = X^1$$

$$X.X = X^2$$

$$X.X.X = X^3$$

$$\sqrt{X}.\sqrt{X} = X$$

$$\sqrt{X} = X^{0.5} \text{ or } X^{\frac{1}{2}}$$

$$\sqrt[3]{X} = X^{0.333} \text{ or } X^{\frac{1}{3}}$$

$$\frac{1}{X^n} = X^{-n}$$

$$X^n.X^m = X^{n+m}$$

$$\frac{X^n}{X^m} = X^{n-m}$$

$$(X^n)^m = X^{nm}$$

$$(X^{-n})^m = X^{m/n}$$

$$\frac{X^n}{X^n} = X^0 = 1$$

Returning to primary and secondary dimensions, it should now be apparent that the three fundamental concepts of mass, length and time are capable of generating many more complex dimensions. They are not sufficient, however, to produce all physical measurements. Thermal measurements are an example. While the concept of heat has the dimensions of energy $[ML^2T^{-2}]$, temperature measures a property not so easily expressed in terms of mass, length and time. Temperature, therefore, is usually regarded as another primary dimension with the symbol $[\theta]$, to denote a fourth property of matter independent of the properties of mass, length and time. Table 2 lists various quantities used in physical geography with their dimensions in terms of the four fundamental dimensions [M], [L], [T], and $[\theta]$.

It is important to realise, even at this early stage, that there is nothing inherently fundamental or unique about the primary dimensions mass, length, time and temperature. Another set of fundamentals could equally well be chosen, provided that all the secondary dimensions are derived consistently from them. Engineers, for example, sometimes prefer to treat force as a fundamental, primary measure, and mass as a quantity derived from the interaction of force, length and time. Denoting the primary dimension of force by [F], the dimensions of mass are defined by force divided by acceleration $[FL^{-1}T^2]$. It is not difficult to redefine all the variables in mechanics according to an [F], [L], [T] system. Pressure, for example, is $[FL^{-2}]$ and energy is [FL]. In the realm of electricity and magnetism, none of the dimensions introduced so far is capable of measuring the properties of electric current or charge, so electric current is often given the status of another primary dimension [I]. Alternatively the electric charge might be defined as the fundamental dimension [Q], in which case the current becomes a secondary quantity. As all definitions of quantities are relative to each other, it does not really matter which quantities are thought of as the starting points. For almost all purposes in physical geography, though, the system of dimensions based on mass, length, time and temperature is convenient and perfectly serviceable.

1.3 Dimensionless numbers

Table 2 is not a complete list of all the quantities likely to be encountered in physical geography, but it can be used as a basis for deriving the dimensions of other variables. When the dimension is written as [1] it means that the quantity has no dimensions. This situation arises when a quantity with dimensions is divided by another quantity having precisely the same dimensions. The result is a pure number, with no units at all. Slope, defined as the ratio of height to length, is a typical example. Height divided by length is written dimensionally as:

$$\frac{[L]}{[L]} = [L][L^{-1}] = [L^0] = [1]$$

Numbers without dimensions are known as dimensionless numbers. Their important property is that they are completely unaffected by changes in units. A slope of 0.32 remains at 0.32, whether the height and length are measured in kilometres, miles, centimetres, fathoms, chains or any other unit of length. Angles when measured in radians (length of arc divided by radius) are dimensionless, and so is strain (also the ratio of two lengths). The constant π ($\approx$3.1416) is another ratio of two lengths (circle circumference and diameter), so π has no units. All proportions and percentages are

TABLE 2. Dimensions of selected quantities in physical geography

Quantity	Definition	Dimensions
Mass	-	[M]
Length, distance	-	[L]
Time	-	[T]
Temperature	-	[θ]
Counts, proportions, probabilities	-	[1]
Area	length x length	[L^2]
Volume	length x length x length	[L^3]
Angle, slope, latitude	length/length	[1]
Area density	number of objects/area	[L^{-2}]
Drainage density	length of streams/area	[L^{-1}]
Velocity	length/time	[LT^{-1}]
Acceleration	velocity/time	[LT^{-2}]
Angular velocity	angle/time	[T^{-1}]
Frequency	number of events/time	[T^{-1}]
Discharge	volume/time	[L^3T^{-1}]
Density	mass/volume	[ML^{-3}]
Force	mass x acceleration	[MLT^{-2}]
Pressure, stress	force/area	[$ML^{-1}T^{-2}$]
Strain	length/length	[1]
Energy, work	force x length	[ML^2T^{-2}]
Viscosity (dynamic)	force x time/area	[$ML^{-1}T^{-1}$]
Momentum	mass x velocity	[MLT^{-1}]
Heat quantity	energy	[ML^2T^{-2}]
Thermal capacity	heat/temperature	[$ML^2T^{-2}\theta^{-1}$]
Specific heat	heat/temperature x mass	[$L^2T^{-2}\theta^{-1}$]
Albedo	energy/energy	[1]

dimensionless. The proportion of an area that is forested, for example, is defined as the forested area divided by the total area, dimensionally $[L^2]/[L^2] = [1]$. Similarly, all probabilities are dimensionless.

Sometimes the combination of several different measures into a composite measure has the effect of cancelling all units and creating a dimensionless number. A good example of this is the Reynolds number, which is used to measure fluid flow characteristics. The Reynolds number Re is defined as:

$$Re = \frac{\rho VL}{\eta}$$

where ρ is the mass density (dimensionally $[ML^{-3}]$), V is the fluid velocity $[LT^{-1}]$, L is a characteristic length (typically the hydraulic radius) with dimension [L] and η is the coefficient of dynamic viscosity $[ML^{-1}T^{-1}]$. This may sound a very complicated measure, but writing out the dimensions of the component variables shows that all the dimensions cancel perfectly and the Reynolds number is therefore a pure number without units:

$$Re = \frac{[ML^{-3}][LT^{-1}][L]}{[ML^{-1}T^{-1}]} = [1]$$

One of the attractions of the Reynolds number is that it measures the change from laminar fluid flow to turbulent flow, a change that occurs when the Reynolds number is about 1000. There are many more examples of dimensionless numbers made up of combinations of variables with units. A method to identify them will be described later.

The last type of dimensionless number is produced by counting a number of objects or events. The number of first order streams in a drainage basin, for instance, the number of hurricanes recorded or the number of plants in a quadrat are all usually treated as being without units. The number of plants per area is therefore given the dimensions $[L^{-2}]$, the number of hurricanes per time period is $[T^{-1}]$, and so on.

1.4 Dimensions in human geography

So far we have been concerned with the variables of interest in physical geography. But what about variables in human geography: do they have dimensions too? If a dimension is a measurable property, then clearly the answer must be yes, although it will certainly be difficult to keep to a system based entirely on mass, length, time and temperature measurements. Unlike the physical sciences, the social sciences have not developed a coherent, interlocking and internationally agreed set of definitions of the properties which are of interest, so the task of identifying dimensions is more difficult. We can make a useful start, however, by agreeing on a set of fundamental quantities which may then be combined to produce more complex measures.

One of the fundamental properties which human geographers deal with and which cannot be expressed in terms of mass, length, time or temperature, is value. This property, which may be measured in monetary units (pounds, dollars) or non-monetary units (preference scales), denotes the worth or esteem accorded to the object in question. It seems sensible to recognise value as a primary dimension, with the symbol [$]. Having done this, it becomes possible to identify the dimensions of other quantities, like the price of land [$$L^{-2}$] or the cost of transporting a certain amount of material a certain distance [$$M^{-1}L^{-1}$].

Another fundamental property in human geography is population size. Strictly speaking, this is measured by counting a number of objects (in this case people) and could, like all counts, be regarded as a dimensionless number without units. Some geographers might insist that this is the correct way to treat population. However, there are several reasons for defining population as a primary dimension in its own right. The first is that the concept of a thousand people is quite different from the concept of 1000 (the pure number). While the latter is simply a number, the former is a number multiplied by a property. This same argument could, of course, be applied to all counts of objects, but the difference between population size and most other counts used in geography is that several secondary quantities of importance in human geography are derived from population size. These include population density, income per capita, frequency per person, and so on. Treating population as a dimension thus opens up the possibility of defining other interlocking variables. Finally, perhaps the best justification for treating population as a primary dimension is that it works. It is possible to get more information out of an equation in which population size is regarded as a measurable property than an equation in which it is treated as a dimensionless number.

In order to apply the concepts of dimensions to human geography we have added two extra primary dimensions: value [$] and population [N]. Thus far, we are on fairly firm ground. Nobody would suppose, however, that all the properties of interest in human geography could be defined in terms of the system: mass, length, time, temperature, value and population. Other suggestions about independent primary dimensions could certainly be made. Maybe the property of information could be usefully thought of as a fundamental quantity, for example. This takes us into areas of speculation and controversy: an exciting prospect, perhaps, but not one for us to tackle here.

Table 3 lists several variables used in human geography and their dimensions according to the [M], [L], [T], [$], [N] system. Comparison with Table 2 will demonstrate that the human geographer does not have as rich a range of secondary dimensions to work with as the physical geographer. This is partly because research in human geography relies much more heavily on counts, percentages, proportions and probabilities than research in physical geography. Consequently many of the variables commonly encountered in human geography are dimensionless.

1.5 Dimensional homogeneity

We can now return to the problem we started with. How is stream frequency related to drainage density? Once we know the dimensions of the quantities involved, the solution is simple. All that is required is to apply the principle of dimensional homogeneity. This states that all quantities that are added, subtracted or equated must have the same dimensions. What does it mean to add five seconds to 1.8 kilometres, or to subtract two hectares from 2°C? The answer, of course, is nothing: such suggestions are ridiculous. In the same way, a speed of 0.5 kilometres per second can never equal a mass of 0.5 kg. If two quantities are equal, not only must the numbers be the same but the properties (dimensions) being measured must also be the same.

TABLE 3. Dimensions of selected quantities in human geography

Quantity	Definition	Dimensions
Mass	-	[M]
Length	-	[L]
Time	-	[T]
Value	-	[$]
Population	-	[N]
Counts, percentages, probabilities	-	[1]
Area	length x length	[L^2]
Population density	population/area	[NL^{-2}]
Population growth	population/time	[NT^{-1}]
Population potential	population/length	[NL^{-1}]
Frequency of events	number of events/time	[T^{-1}]
Man hours	population x time	[NT]
Income	value/population x time	[$\$N^{-1}T^{-1}$]
Land rent	value/area x time	[$\$L^{-2}T^{-1}$]
Cost per time	value/time	[$\$T^{-1}$]
Transport rate	value/mass x distance	[$\$M^{-1}L^{-1}$]
Crop yield	mass/area x time	[$ML^{-2}T^{-1}$]
Resources per person	mass/population	[MN^{-1}]

If we use $\underline{F}$ to denote channel frequency and $\underline{D}$ for drainage density then we can find out how they must be related using trial and error. First, let us try the equation:

$$F = D$$

Now we replace the variables with their dimensions. Channel frequency is the number of streams (dimensionless) divided by area, or [L^{-2}]. Drainage density is the total length of streams [L] divided by area, which is [L/L^2] or [L^{-1}]. This gives the relationship between $\underline{F}$ and $\underline{D}$ as:

$$[L^{-2}] = [L^{-1}]$$

which cannot be correct as the quantities being equated do not have the same dimensions. Let us try another equation:

$$F = \sqrt{D}$$

Written in dimensions, this is:

$$[L^{-2}] = \sqrt{[L^{-1}]}$$
$$= [L^{-0.5}]$$

which is also nonsense, because the dimensions still do not balance. Attempting something else, suppose

$$F = D^3$$

This would give

$$[L^{-2}] = [L^{-1}]^3$$
$$= [L^{-3}]$$

which is no better. All the equations attempted so far are physically impossible because they try to equate properties that are not the same. Obviously we could continue this game, thinking of all the various ways in which the two quantities might possibly be related and rejecting each one because it did not satisfy the principle of dimensional homogeneity. At last we might try:

$$F = D^2$$

which in terms of dimensions is:

$$[L^{-2}] = [L^{-1}]^2$$
$$= [L^{-2}]$$

This is what we want: an equation in which the property measured on one side is balanced by precisely the same property on the other side. The equation $F = D^2$ is dimensionally homogeneous and therefore it is physically possible. It is the only way in which these two quantities can be equated, because all the other ways fail to satisfy the principle of dimensional homogeneity.

When an equation is dimensionally homogeneous it is physically possible, but it is not necessarily complete. A more correct way to write the relationship would be

$$F \propto D^2$$

In other words, channel frequency is proportional to drainage density squared. To make both sides exactly equal we need a scaling constant on the right hand side. This is to multiply the number D^2, to adjust this number upwards or downwards to make it equal to the number on the left hand side of the relationship. The constant of proportionality is a dimensionless constant, so its presence makes no difference to the dimensional balance of the equation. Representing the dimensionless constant of proportionality by C, the full equation is:

$$F = CD^2$$

What does all this achieve? The answer is an increase in our understanding of river drainage patterns without doing any fieldwork or data analysis. Not that fieldwork and data analysis are to be discouraged; on the contrary, they are indispensable to find the numerical value of C. Dimensional analysis can never reveal the actual number a constant stands for in an equation. It can, however, pinpoint the most likely relationship between a set of variables, and to know this before you start collecting data is a considerable help. As a result of examining dimensions, we now know that if the number of streams in a given area is doubled then drainage density will be quadrupled, because the equation is only possible when F is proportional to D^2. You will be reassured to learn that this law was 'discovered' the hard way by Melton (1958) after a painstaking study of 156 drainage basins. Melton's empirical data suggested a value of 0.694 for C.

2. CHECKING EQUATIONS AND UNITS

The principle of dimensional homogeneity is a useful rule for checking equations in geography. An equation which does not balance dimensionally has something wrong with it. Of course, dimensional balance alone does not make an equation true. After all, to say

Distance from London to Bristol = Distance from Stockholm to Uppsala

is not true, yet the equation does satisfy the dimensional homogeneity principle. You will notice, however, that the equation becomes true if one of the distances is multiplied by a dimensionless constant of proportionality.

In this section we examine some typical geographical equations in order to identify the dimensions of the quantities they contain and to verify that they are balanced. After that, we shall use the same method to find errors in equations.

2.1 Some geographical equations

One of the most common forms of equation in geography is the regression equation of statistics, which expresses a dependent variable Y as a linear function of an independent variable X and two constants a and b:

$$Y = a + bX$$

Suppose a regression analysis is conducted to find out the relationship between the midday temperature Y measured in degrees Centigrade and altitude X, measured in metres. The dimensions of Y would then be $[\theta]$ and those of X would be $[L]$. The intercept constant a (which gives the temperature at zero altitude) must have the dimensions of temperature $[\theta]$. The regression coefficient b gives the change in temperature associated with an increase of one unit of altitude. Since it expresses changes in temperature relative to changes in altitude, it must have the dimensions $[\theta L^{-1}]$. Written in dimensions, this particular regression equation is therefore:

$$[\theta] = [\theta] + [\theta L^{-1}][L]$$

In the term $[\theta L^{-1}][L]$ a temperature is divided by a length and then multiplied by a length. The lengths therefore cancel each other out and temperature is left, so the equation is really

$$[\theta] = [\theta] + [\theta]$$

A temperature equals a temperature plus a temperature: the principle of dimensional homogeneity is satisfied.

Regression equations using logarithms need slightly different treatment. How can a logarithm have a dimension? The answer is that logarithms do not have dimensions: they are pure numbers. To write a logarithmic regression equation in terms of its dimensions, the variables must be transformed out of logarithms. Take for example the non-linear regression equation:

$$\log Y = \log a + b \log X$$

where log Y is the dependent variable, log X is the independent variable, log a is the intercept constant and b (the regression coefficient) is the slope of the regression line. Out of logarithms, the equation is:

$$Y = aX^b$$

The dimensions of the original untransformed variables can be entered into this equation without any difficulty. To satisfy the principle of dimensional homogeneity, the constant a must have the dimensions of (Y/X^b). This makes sense when you remember that the literal interpretation of the constant a is the value of Y when $X^b = 1$. The exponent b is dimensionless.

Another type of equation which is common in geography is the exponential function, such as:

$$P = P_o e^{bt}$$

which describes the exponential growth of a population. Here P is the population at time t, P_o is the population at time t = 0, and b is a constant which gives the proportional increase in population per unit of time (dimensionally $[T^{-1}]$). The constant e (the base of the natural logarithms) is dimensionless. Inserting the dimensions of the quantities, the equation may be seen to be perfectly balanced:

$$[N] = [N][1][T^{-1}][T]$$

A more complicated exponential function is exemplified by the family of spatial interaction models. Consider a shopping trip distribution model of the form:

$$T_{ij} = A_i O_i D_j \; e^{-\beta c_{ij}}$$

where the terms and their dimensions are as follows:

T_{ij}	number of shopping trips from zone i to zone j per time	$[T^{-1}]$
O_i	total number of shopping trips from zone i per time	$[T^{-1}]$
D_j	amount of retail floorspace in zone j	$[L^2]$
c_{ij}	cost of travelling from zone i to zone j	[$]
β	(constant) proportional trip reduction per unit of cost	$[\$^{-1}]$
e	(constant) base of the natural logs	[1]
A_i	(constant) ratio between O_i and the sum of the unscaled predicted trips from zone i	$[L^{-2}]$

The constant A_i is simply a balancing factor that makes sure the total number of predicted trips from i are the same as O_i, which is known beforehand. It is defined as:

$$A_i = \frac{1}{\sum_j D_j e^{-\beta c_{ij}}}$$

which, because the term $e^{-\beta c_{ij}}$ is dimensionless, has the dimensions of $1/D_j$, or (L^{-2}). The dimensions of the complete model can therefore be written:

$$[T^{-1}] = [L^{-2}][T^{-1}][L^2][1][\$^{-1}][\$]$$

which balances.

2.2 The dimensions of constants

Once you can check that equations are dimensionally balanced, it is not difficult to use dimensional analysis as an aid to understanding relationships. You can discover the dimensions of quantities which have been accidentally left out of an equation, or work out the likely definition of a variable or a constant simply from its position in an equation.

Newton's law of gravitation is the first illustrative example. The force F of attraction between two bodies is proportional to the mass of the first body m_1 multiplied by the mass of the second body m_2 and inversely proportional to the square of the distance r separating them. This could be written as an equation:

$$F = \frac{m_1 m_2}{r^2}$$

Substituting the dimensions of the variables:

$$[MLT^{-2}] = \frac{[M][M]}{[L]^2}$$

This equation does not balance. Something must be missing on the right side of the equation which, if it were present, would make the dimensions balance. Otherwise, Newton's law would be physically impossible. Adding the mystery ingredient X to the equation:

$$F = \frac{m_1 m_2}{r^2} X$$

The dimensions of X are not yet known, so we can represent them only by a question mark in the dimensional equation:

$$[MLT^{-2}] = \frac{[M][M]}{[L]^2} [?]$$

Taking all the known dimensions to the left side enables us to identify the nature of the mystery ingredient:

$$[MLT^{-2}]\frac{[L]^2}{[M][M]} = [?]$$

$$[MLT^{-2}]\ [L]^2\ [M]^{-1}\ [M]^{-1} = [?]$$

$$[L^3M^{-1}T^{-2}] = [?]$$

This reasoning leads to the conclusion that a quantity measured in dimensions of $[L^3M^{-1}T^{-2}]$ has been left out of the gravitation equation. In fact, what has been left out is the universal gravitation constant (G), which is equal to 6.673 x 10^{-11} m^3 kg^{-1} s^{-2}, so its dimensions are indeed $[L^3M^{-1}T^{-2}]$. The full balanced equation is:

$$F = G\ \frac{m_1 m_2}{r^2}$$

Taking another example, the gas law which governs the behaviour of the atmosphere may be written as:

$$p = R\rho T$$

where p is the pressure of the atmosphere, ρ is the density of the atmosphere, T is the temperature and R is a gas constant. The units of measurement of R are not immediately obvious, to say the least, but they can be deduced easily enough from the dimensions of the other quantities in the equation. Pressure has the dimension of force per unit area $[MLT^{-2}][L^{-2}]$ or $[ML^{-1}T^{-2}]$, density is $[ML^{-3}]$ and temperature can itself be regarded as a primary dimension $[\theta]$:

$$[ML^{-1}T^{-2}] = R[ML^{-3}][\theta].$$

Rearranging the equation and cancelling gives:

$$R = [L^2T^{-2}\theta^{-1}]$$

These are very odd dimensions indeed. They show that the numerical value of the gas constant is very much influenced by whatever units of length, time and temperature are chosen to measure the variables in question. Changing the units of length from metres to centimetres, for example, would increase the value of R not by 100 but by 10 000 (since the length dimension is squared). On the other hand, altering the unit of mass from kilograms to grams would have no effect at all on R. We shall return to these problems of changing units shortly.

2.3 Errors in equations

Once you are conscious of units of measurement and dimensions as they appear in equations it becomes easier to spot other people's mistakes. The following illustration is taken from a climatology textbook which uses an equation for centripetal acceleration in its explanation of the direction of pressure gradient winds. The equation is

$$c = -\frac{mv^2}{r}$$

where c is the centripetal acceleration, m is the mass of an object that is moving in a curved path, v is its velocity and r is the radius of curvature. Substituting the dimensions of the variables into the equation:

$$[LT^{-2}] = \frac{[M][LT^{-1}]^2}{[L]}$$

This is not an equation at all, since the left hand side is obviously not equal to the right hand side. One of two things could be wrong: either the mass m should be left out of the expression altogether, so allowing it to balance, or else one of the variables has been wrongly defined. In fact c has been wrongly defined. It is really the centripetal force, and not an acceleration at all. When c is given the dimensions of force $[MLT^{-2}]$ then equilibrium is restored. Not many textbooks in geography contain definitions or equations that are obviously wrong, but it is always useful to have a quick method with which to investigate anything that looks suspicious.

Here is another example, which appeared in a research paper in economic geography. It concerns the costs of transporting goods in relation to the distance travelled. With some slight rearranging, the equation given was:

$$y^2 = D_1{}^2r^2 + 2\,D_1D_2r - s$$

where the variables were defined as:

y	total costs	[\$]
D_1	variable costs per mile	$[\$L^{-1}]$
D_2	fixed overhead charges	[\$]
r	distance	[L]
s	initial outlays for equipment and administrative expenses	[\$]

To this list, I have added the dimensions of the quantities, using the dimension [\$] to denote money. In dimensional form the equation is:

$$[\$]^2 = [\$L^{-1}]^2[L]^2 + [1][\$L^{-1}][\$][L] - [\$]$$

When the length dimensions are cancelled,

$$[\$]^2 = [\$]^2 + [\$]^2 - [\$]$$

The last term is clearly not correct. It is no more possible to subtract a cost from a cost squared than it is to speak of an area minus a length. Two solutions to the difficulty might be suggested. Either s, the initial outlays for equipment and administrative expenses, should be squared in the formula or, alternatively, s should be multiplied by another cost.

2.4 Empirical relationships

What about equations which are not dimensionally balanced, but which appear to describe what happens in the real world? The geographical literature is full of empirical relationships which 'work' but which have no dimensional justification. Examples might be relationships between crime rates and climatic conditions, between the number of shoppers and the perceived attractiveness of the shopping centre, or between the number of species present and the altitude of a study area. Typically, equations are the result of regression analysis, often on the logarithms of variables which automatically leads to a power function formulation. For instance, if crime rates were related to mean July temperature, the regression equation might look like this:

$$\log R = 0.06 + 2.4 \log J$$

where R is the crime rate per thousand population per month and J is the mean July temperature. Out of logarithms this becomes:

$$R = k\ J^{2.4}$$

where k is a constant equal to the antilogarithm of 0.06. Giving R the dimensions $[N^{-1}T^{-1}]$ (number of crimes per thousand population per month) and J the dimension $[\theta]$, the equation in dimensions is:

$$[N^{-1}T^{-1}] = k[\theta]^{2.4}$$

which clearly does not balance. One way out is to give the constant k the dimensions necessary to make the equation homogeneous, in this case the rather unlikely combination $[N^{-1}T^{-1}\theta^{-2.4}]$ so that:

$$[N^{-1}T^{-1}] = [N^{-1}T^{-1}\theta^{-2.4}][\theta]^{2.4}$$

But this is merely to paper over the cracks; to cover up a gaping hole of ignorance. Nobody knows what such a constant would actually mean. Without denying that it would be interesting and perhaps even useful to be able to predict crime rates from July temperatures, it remains true that we still have no idea why the equation might match reality and under what circumstances it might not. In other words, we have no theory, and several variables which come between July temperatures and crime rates have been omitted. An empirical equation might be a step towards understanding, but it is far from complete if it does not make dimensional sense. However well an empirical equation seems to fit the real world, if it is dimensionally unbalanced this is a clear signal that something is wrong.

2.5 Which way round?

Finding the remedy for the deficiencies of empirical relationships is a difficult and rather rarified task. Dimensional analysis has much simpler and more mundane applications which the geographer will find useful. It is fairly common, for instance, for people to get muddled over an equation in the heat of the moment. When this happens, the problem can usually be resolved by thinking about the dimensions involved. For example, the discharge of a stream is measured by multiplying the velocity of the stream flow by the cross-sectional area of the channel. Or is it that the discharge is measured by dividing the velocity by the cross-sectional area? If you are not sure, write down the dimensions of the alternatives. Discharge is volume of water per time $[L^3T^{-1}]$, velocity is length per time $[LT^{-1}]$ and area is $[L^2]$, so the alternatives are:

Alternative 1 $$[L^3T^{-1}] = [LT^{-1}][L^2]$$

Alternative 2 $$[L^3T^{-1}] = \frac{[LT^{-1}]}{[L^2]}$$

There can now be no doubt that the first alternative is correct.

2.6 Changing units

The last example also shows the usefulness of reducing a variable to its dimensions when it is required to change from one system of units to another. Discharge is often measured in cubic feet per second, but suppose it is necessary to change to SI (Système International) units where the metre is the basic measure of length. What is a discharge of 230 cubic feet per second in SI units? There are 0.3048 metres in 1 foot, but this conversion factor must be raised to the power of 3 (like the length dimension it modifies) to make the change:

$$\begin{aligned} 230\ ft^3\ s^{-1} &= 230 \times (0.3048)^3 \\ &= 6.51\ m^3\ s^{-1} \end{aligned}$$

When more than one unit has to be changed, the same rule applies. Each conversion factor is raised to the same power as the primary dimension to which it refers. To convert the density of water in SI units ($1000\ kg\ m^{-3}$) into pounds per cubic foot needs two conversion factors:

$$1\ kg = 2.205\ lb$$
$$1\ m = 3.281\ ft$$

As the dimensions of density are $[ML^{-3}]$, the first conversion factor is applied without alteration (raised to the power of 1), while the second is raised to the power of -3:

$$\begin{aligned} 1000\ kg\ m^{-3} &= 1000 \times (2.205)^1 \times (3.281)^{-3} \\ &= 62.43\ lb\ ft^{-3} \end{aligned}$$

In this way conversions between units can be kept simple.

3. DERIVING RELATIONSHIPS

So far we have seen that dimensional analysis can be used for checking equations, either to make sure they are correctly given or to understand how the quantities are to be measured. But we can go further than this. Dimensional analysis can also be used to predict relationships in advance. Provided that we know which variables are relevant to the problem in question, there is often only one way in which they can be related in a power function form. If we can find a way of combining variables into a dimensionally homogeneous equation, then this will reveal the relationship before any experiment is attempted. An empirical study can then be made to confirm the theoretically expected equation and to fill in any gaps.

3.1 The basic method

Earlier we considered the relationship between channel frequency and drainage density. Using a trial and error method it was shown that several ways of relating the two quantities in an equation were not physically possible because they were not dimensionally homogeneous. One dimensionally homogeneous equation was identified, and while it was asserted that this was the only one possible, the assertion was not proved. Now we shall prove it, and demonstrate a more elegant short cut method which will replace trial and error.

The relevant quantities, with their dimensions are:

F channel frequency $[L^{-2}]$

D drainage density $[L^{-1}]$

If F is related to D in a power function of an unknown nature, then we can tentatively write the equation as:

$$F = C\,D^{a}$$

where C is an unknown dimensionless constant (included in case there is a difference in proportionality between the two sides of the equation) and a is an unknown exponent. If we have no idea what the relationship between the quantities is, then the value of a could be anything. The task is therefore to find out exactly what a is. First, we write the same equation in terms of dimensions:

$$[L^{-2}] = [1][L^{-1}]^{a}$$

where C is represented by its dimensionless status [1]. Now, we know from the principle of dimensional homogeneity that the dimensions on the left hand side must be exactly balanced by the dimensions on the right side. The dimension [L] has an exponent of -2 on the left, so it must also have an exponent of -2 on the right. But we already know that the exponent of [L] on the right is -1a. It follows that, to balance the exponents of [L] on both sides,

$$-2 = -1a$$

If -a is -2, then a must be equal to 2. In fact, a cannot be equal to anything but 2. This proves that there is only one way in which these particular quantities can be combined in a dimensionally balanced equation. Substituting 2 for a, we are left with the unique solution:

$$F = C\,D^2$$

The only thing missing is that we still do not know the numerical value of C, the dimensionless constant, which must be discovered by examining empirical data. But, apart from that, dimensional analysis has derived the complete relationship. Most equations in physics which were discovered after painstaking experiments and complex theoretical arguments can in fact be derived through dimensional analysis in a few moments. The same is true of many equations used in geography, as the examples following will show.

3.2 Waves in shallow water *

The speed of waves near a coast has been observed to vary with the depth of the water. As the water gets shallower, the waves are retarded. As the waves are subject to the influence of gravity, the acceleration due to gravity is also expected to form part of the relationship. The quantities are:

v	wave velocity	$[LT^{-1}]$
g	acceleration due to gravity	$[LT^{-2}]$
h	depth of water	$[L]$

Note that this example is more complicated than the last one: we are now relating three, rather than two, quantities. If we write wave velocity as an unknown power function of g and h (with C a dimensionless constant):

$$v = C\,g^a\,h^b$$

where both a and b are exponents whose numerical values we need to find out. In dimensions this becomes:

$$[LT^{-1}] = [1][LT^{-2}]^a[L]^b$$

The principle is the same as before: the dimensions on both sides of the equation must balance. First, consider the dimension [L]. If we ignore everything else but the length dimension, the equation looks like this:

$$[L] = [L]^a[L]^b$$

The length on the left side can be written as length raised to the power of 1:

$$[L]^1 = [L]^a[L]^b$$

What has to be done now is to make sure that the exponents of [L] balance on both sides of the equation, so that

$$1 = a + b \qquad \text{(i)}$$

Keeping this in mind, we can turn to the dimension [T]. Ignoring everything else, the [T] part of the power function equation is:

$$[T^{-1}] = [T^{-2}]^a$$

To keep the dimension [T] the same on both sides of the equation, the exponent of [T] on the left (-1) must equal the exponent (-2a) on the right:

$$-1 = -2a \qquad \text{(ii)}$$

Equations (i) and (ii) referring to the exponents of [L] and [T] respectively must be true simultaneously if the relationship is to be dimensionally homogeneous. Using the information available from both simultaneous equations the values of a and b can be found. From (ii) $a = \frac{1}{2}$, so if $\frac{1}{2}$ is substituted for a in equation (i) we see that $b = \frac{1}{2}$. The dimensionally balanced equation linking wave velocity, gravity and water depth is therefore:

$$v = C\, g^{\frac{1}{2}}\, h^{\frac{1}{2}}$$

or

$$v = C\sqrt{gh}$$

The velocity of a wave in shallow water varies with the square root of the water depth. Dimensional analysis has revealed that as waves approach a coastline up a gently shelving beach their speed is reduced at a greater rate than the reduction in depth. The result of this process is that waves that are evenly spaced out to sea are compressed together nearer the beach as their velocity is greatly reduced. Dimensional analysis reveals nothing about the constant of proportionality C, whose numerical value must be found by experimental measurement. A simple experiment relating the variables would discover that the constant is, in fact, unity, so C disappears from the equation, to give

$$v = \sqrt{gh}$$

3.3 Velocity of seismic waves

Water is not the only medium which generates waves. The solid Earth itself transmits oscillating shock waves created by deep-seated movements. The waves generated by earthquakes can be divided into Primary or P-waves, which reach a seismograph first because they travel faster than the Secondary or S-waves. There are also L-waves, which travel at the surface, unlike P and S-waves which travel through the Earth. P-waves are compressional or 'push' waves in which the motion of the material is in the direction of the wave. S-waves are 'shear' waves, in which the motion of material is transverse to the direction of the wave. The speed of P and S waves passing through rock is determined by variables quite different from those affecting water waves. Seismic wave velocity depends firstly on the density of the rock and secondly on the compressibility (or incompressibility) of the rock. The incompressibility of the rock is measured in the direction of the wave by the axial modulus for P-waves, and at right angles to the direction of the wave by the rigidity modulus for S-waves. Apart from the differences of direction, the speed of P and S-waves is governed by very similar relationships. We shall concentrate here on P-waves.

The axial modulus expresses the ratio between a force per unit area (stress) applied to the rock and the degree of deformation resulting (strain). Force per unit area is dimensionally $[MLT^{-2}]/[L^2]$ or $[ML^{-1}T^{-2}]$. Strain is simply the ratio of two lengths: the change in length resulting from the stress divided by the original length. Strain is therefore dimensionless, and the axial modulus has the dimensions $[ML^{-1}T^{-2}]$:

$$\text{axial modulus} = \frac{\text{longitudinal stress}}{\text{longitudinal strain}} = \frac{[ML^{-1}T^{-2}]}{[1]} = [ML^{-1}T^{-2}]$$

To express the velocity of a P-wave as a function of rock density and axial modulus, we first set out the variables:

V_p	velocity of P-wave	$[LT^{-1}]$
ρ	density of rock	$[ML^{-3}]$
ψ	axial modulus of rock	$[ML^{-1}T^{-2}]$

Expressing wave velocity as a power function of density and axial modulus:

$$V_p = C(\rho)^a(\psi)^b$$

where C is a dimensionless scaling constant. Written dimensionally:

$$[LT^{-1}] = [1][ML^{-3}]^a\ [ML^{-1}T^{-2}]^b$$

The exponents a and b each have only one value that will enable the dimensions on both sides of the equation to balance.

Balancing the exponents of	[L]	$1 = -3a - b$
" " " "	[T]	$-1 = -2b$
" " " "	[M]	$0 = a + b$

(Note that the dimension [M] does not appear in V_p, so we say the exponent of [M] is zero when V_p is written dimensionally.) The three simultaneous equations balancing the exponents can now be solved to find the values of a and b. From the second equation, $b = \frac{1}{2}$, and so $a = -\frac{1}{2}$ from either of the other two equations. Inserting these numerical values into the power function,

$$V_p = C\rho^{-\frac{1}{2}}\psi^{\frac{1}{2}},$$

which is the same as

$$V_p = C\sqrt{\frac{\psi}{\rho}}$$

This is the only way in which rock density and rock incompressibility (measured by axial modulus) can be related to P-wave velocity without upsetting the dimensional balance. The constant of proportionality actually has a value of 1 when a consistent system of units is used to measure the three variables, but that is discovered from experiment, not from dimensional analysis.

3.4 The barometer equation

The barometer is an instrument which measures air pressure from the height of a column of liquid in a narrow vertical tube. The column of liquid is balanced so that if the air pressure increases the liquid is pushed higher up the tube. If air pressure falls, the level of liquid drops in compensation. The air pressure can therefore be measured as a function of the height of the column of liquid. Constants which are expected to influence the relationship are the acceleration due to gravity, since the liquid drains out of the tube under the influence of gravity when air pressure is reduced, and the density of the liquid. A simple barometer could be constructed using any liquid, but mercury is usually chosen because of its high density: just a few centimetres of mercury is enough to balance the weight of thousands of metres of the atmosphere. A liquid with a lower density would rise much higher up the tube than mercury for the same atmospheric pressure. The variables and their dimensions can therefore be set out as follows:

p	air pressure	$[ML^{-1}T^{-2}]$
g	acceleration due to gravity	$[LT^{-2}]$
ρ	density of liquid	$[ML^{-3}]$
h	height of liquid column	$[L]$

Writing air pressure as a function of the other three quantities:

$$p = C\, g^a\, \rho^b\, h^c,$$

where C is a dimensionless constant. In terms of dimensions:

$$[ML^{-1}T^{-2}] = [1]\, [LT^{-2}]^a\, [ML^{-3}]^b\, [L]^c$$

Solving for mass: $1 = b$
" " length: $-1 = a - 3b + c$
" " time: $-2 = -2a$

So, a = 1, b = 1 and c = 1

The equation relating air pressure to the height of a column of liquid, the density of the liquid and gravity is therefore:

$$p = Cg\rho h$$

3.5 The geostrophic wind equation

If the Earth did not rotate, winds would follow the pressure gradient force from high to low pressure areas. Rotation produces another force, the Coriolis force, which deflects moving air to the right of its path in the northern hemisphere and to the left in the southern hemisphere. The Coriolis force is absent at the equator and strongest near the poles. In the free atmosphere the two forces reach equilibrium when the wind has been turned so that it flows in a direction at right angles to the pressure gradient. This resultant wind at high altitudes is called the geostrophic wind.

The velocity of the geostrophic wind varies with the horizontal pressure gradient, the strength of the Coriolis force (a function of latitude) and the density of the atmosphere (a function of altitude). The horizontal pressure gradient is the change in pressure (p) with horizontal distance (n) at that point. It is expressed as the partial derivative of p with respect to n:

$$\text{horizontal pressure gradient} = \frac{\delta p}{\delta n} = \frac{[ML^{-1}T^{-2}]}{[L]} = [ML^{-2}T^{-2}]$$

Although we have had to resort to partial differential calculus to express this rate of change in pressure, the dimensions involved are straightforward: pressure divided by length, to give $[ML^{-2}T^{-2}]$. In general there is no need to be put off by the symbols of calculus when reducing a variable to its dimensions. Differentials, partial differentials and integral signs just melt away to leave the dimensions of the quantities involved.

The strength of the Coriolis force at a particular latitude is best expressed by the Coriolis parameter (f). This is equal to ($2\Omega \sin \phi$), where Ω is the Earth's angular velocity measured in radians per second and ϕ is the latitude. Since radians and latitude are both dimensionless, the dimensions of the Coriolis parameter reduce to $[T^{-1}]$:

$$\text{Coriolis parameter} = f = [T^{-1}]$$

The final variable, ρ, which is the density of the atmosphere at a certain altitude has the dimensions $[ML^{-3}]$.

Expressing the velocity of the geostrophic wind (V) as a power function of the three quantities, with the addition of a dimensionless constant C:

$$V = C\, f^a\, \rho^b \left(\frac{\delta p}{\delta n}\right)^c$$

The dimensional equation is

$$[LT^{-1}] = [1]\ [T^{-1}]^a\ [ML^{-3}]^b\ [ML^{-2}T^{-2}]^c$$

Solving for [L]: $1 = -3b - 2c$

" " [M]: $0 = b + c$

" " [T]: $-1 = -a - 2c$

The solution is $a = -1$, $b = -1$, $c = 1$, so the final equation for the velocity of the geostrophic wind is:

$$V = \frac{C}{f\rho}\left(\frac{\delta p}{\delta n}\right)$$

3.6 Central place theory example

To demonstrate that dimensional analysis can be used to derive relationships in human geography as well as those describing the physical environment, the next example is related to central place theory. On an isotropic plain, where population is evenly distributed and shops are optimally located so that people spend as little as possible on travel, what determines the maximum distance anyone has to travel to obtain a certain good? The maximum distance depends on the level of sales per time period necessary to support a shop which provides the good (in other words, the threshold of the good). It also depends on the average spending per person per time on the good, and on the population density. Denoting population by the dimension [N] the variables may be written:

r	maximum travel distance	[L]
t	threshold of good	$[\$T^{-1}]$
y	average spending on good	$[\$N^{-1}T^{-1}]$
w	population density	$[NL^{-2}]$

The power function relationship between the variables must be of the general form:

$$r = C\ t^a\ y^b\ w^c$$

where C is a dimensionless constant. In dimensions:

$$[L] = [1]\ [\$T^{-1}]^a\ [\$N^{-1}T^{-1}]^b\ [NL^{-2}]^c$$

Balancing [L]: $1 = -2c$

" [\$]: $0 = a + b$

" [T]: $0 = -a - b$

" [N]: $0 = -b + c$

From these simultaneous equations we see that $a = \frac{1}{2}$, $b = -\frac{1}{2}$, and $c = -\frac{1}{2}$. Raising a quantity to the power $\frac{1}{2}$ is the same as taking the square root, so the relationship between the variables is:

$$r = C\sqrt{\frac{t}{yw}}$$

The maximum distance to obtain a good is therefore expected to respond less than proportionally to variations in population density or demand. Doubling spending - or doubling the population density - would bring more shops into the region and so reduce travelling distances, but not by as much as half. The constant C would vary in the real world according to how much overlap existed between shops' trade areas. In the perfect Christaller system where trade areas are non-overlapping hexagons the term t/yw would give the area of each hexagon. You can see that t/yw is an area by examining the combined dimensions:

$$\frac{[\$T^{-1}]}{[\$N^{-1}T^{-1}][NL^{-2}]}$$

The [\$], [T] and [N] dimensions all cancel, and $[L^{-2}]$ in the denominator becomes $[L^2]$, or area. In a Christaller landscape, the maximum travel distance r is the distance between the centre of a hexagon and one of its corners, so the constant C becomes the ratio of this distance to the square root of hexagon area. The ratio is 0.62 for any regular hexagon. Values of C greater than 0.62 in the real world would result from deviations from Christaller's ideal.

While this example may not reveal more than you knew already about central place theory, it does show that checking the logic of a theory and working out its mathematical implications is facilitated by an understanding of dimensions. Given the scarcity of theory in human geography it is unfortunately true that human geographers do not have as many opportunities for this type of analysis as physical geographers.

3.7 Spurious relationships

Under some circumstances, dimensional analysis is discriminating enough to reject a variable which has been wrongly included in the list of relevant quantities. To take the central place theory example a step further, suppose that the 'transportability' of a good was thought to influence the maximum distance people would travel for it under isotropic conditions. Transportability might be measured as the weight (mass) of the good compared with its value, or mass per unit value $[M\$^{-1}]$. Let us add this new variable to the list and proceed in the normal way:

r	maximum travel distance	$[L]$
t	threshold of good	$[\$T^{-1}]$
y	average spending on good	$[\$N^{-1}T^{-1}]$
w	population density	$[NL^{-2}]$
h	transportability of good	$[M\$^{-1}]$

$$r = C\, t^a\, y^b\, w^c\, h^d$$

$$[L] = [1]\, [\$T^{-1}]^a\, [\$N^{-1}T^{-1}]^b\, [NL^{-2}]^c\, [M\$^{-1}]^d$$

Balancing	$[L]$:	$1 = -2c$
"	$[\$]$:	$0 = a + b - d$
"	$[T]$:	$0 = -a - b$
"	$[N]$:	$0 = -b + c$
"	$[M]$:	$0 = d$

The solutions are $a = \frac{1}{2}$, $b = -\frac{1}{2}$, $c = -\frac{1}{2}$ and $d = 0$. The transportability h is therefore raised to the power of zero and disappears from the relationship, which remains:

$$r = C\sqrt{\frac{t}{yw}}$$

The method cannot always be relied upon to exclude irrelevant variables, however. Although the derivation of relationships by dimensional analysis might seem like magic, it only works when all, or most, of the included variables are known to be relevant to the problem being considered. It is quite possible to draw up a list of variables which in fact are not functionally related together at all, and still arrive at a respectable-looking (but quite spurious) derived relationship using dimensional analysis. Suppose, for example, that it is postulated without any theoretical justification at all that the price of land is a function of population density, average income and the rate of inflation:

P	land price	$[\$L^{-2}]$
w	population density	$[NL^{-2}]$
I	average income	$[\$N^{-1}T^{-1}]$
R	inflation rate (annual percentage)	$[T^{-1}]$

With the usual method, a relationship is derived:

$$P = C \frac{wI}{R}$$

This equation has its dimensions perfectly balanced, but it is geographical nonsense. Dimensional homogeneity is a necessary but not a sufficient condition for an equation to be acceptable. Dimensional analysis is not an easy trick which reveals nature's secrets effortlessly. It is effective only after an accurate identification of relevant variables, which must either be done theoretically or by experiment.

4. PARTIAL SOLUTIONS

Dimensional analysis is most satisfying when used to prove that only one relationship is possible between a number of variables. Complete solutions giving the numerical values of exponents in power function equations are not uncommon, as was demonstrated in the preceding section. Not all attempts to derive relationships end so neatly, however. When there are more independent variables than there are simultaneous equations only a partial solution can be achieved. In this section the method of partial solution will be illustrated and a tactic for avoiding partial solutions in some special circumstances will be outlined. The first examples will show how to work out the algebra for a partial solution.

4.1 Traffic flow

Consider the relationship between traffic flow (vehicles per hour) and four other variables:

F	traffic flow rate	$[T^{-1}]$
u	average traffic speed	$[LT^{-1}]$
h	road length per area	$[L^{-1}]$
v	vehicles per 100 population	$[N^{-1}]$
w	population density	$[NL^{-2}]$

Arranging the variables as a power function, and then in terms of their dimensions:

$$F = C\ u^a\ h^b\ v^c\ w^d$$

$$[T^{-1}] = [1]\ [LT^{-1}]^a\ [L^{-1}]^b\ [N^{-1}]^c\ [NL^{-2}]^d$$

Balancing the exponents of each primary dimension gives three simultaneous equations:

For $[T]$: $-1 = -a$

$[L]$: $0 = a - b - 2d$

$[N]$: $0 = -c + d$

Three equations containing four unknowns are impossible to solve completely. From the first equation $\underline{a} = 1$. Substituting 1 for $\underline{a}$, the second equation becomes:

$$0 = 1 - b - 2d$$

so $\underline{b} = 1 - 2\underline{d}$. It is already obvious from the third equation that $\underline{c} = \underline{d}$, so we at least are able to simplify $\underline{b}$ and $\underline{c}$ by expressing them in terms of $\underline{d}$. Rewriting the power function with the partial solutions inserted:

$$F = C\, u^1\, h^{1-2d}\, v^d\, w^d$$

Since:

$$h^{1-2d} = h^1\left(\frac{1}{h^{2d}}\right)$$

the partial solution is:

$$F = C\, uh\left(\frac{vw}{h^2}\right)^d$$

The variable $\underline{h}$ now appears twice in the expression, but the advantage of writing the equation like this is that only one unknown exponent remains to be evaluated. The relationship has been partially, but not fully, solved.

4.2 The orbit of a planet

All the planets revolve around the sun and are held in their orbits by the forces of gravity. The time it takes a planet to complete one revolution of the sun is expected to be a function of the strength of the force of attraction between the planet and the sun. We know from Newton's gravity formula, which we discussed earlier, that the attractive force is itself a function of the two masses, the distance between them and the universal gravitational constant. Can we express the periodic time of the orbit in terms of these quantities? Let us try to do this, remembering that the distance between the planet and the sun varies somewhat since the orbit is elliptical. To overcome this difficulty, the distance measurement is made along the long (or major) axis of the orbit. The various quantities are set out in the usual way:

t	periodic time	[T]
m_s	mass of sun	[M]
m_p	mass of planet	[M]
r	major axis of orbit	[L]
G	universal gravitational constant	$[L^3M^{-1}T^{-2}]$

Writing periodic time as a power function of the other quantities (with the dimensionless parameter $\underline{C}$):

$$t = C\, m_s^a\, m_p^b\, r^c\, G^d$$

$$[T] = [1]\,[M]^a\,[M]^b\,[L]^c\,[L^3M^{-1}T^{-2}]^d$$

Balancing [T]: $1 = -2d$

" [L]: $0 = c + 3d$

" [M]: $0 = a + b - d$

It is clear from the first two simultaneous equations that $d = -\frac{1}{2}$ and $c = {}^3{}_2$, but the only information about the values of a and b is that:

$$a + b = -\tfrac{1}{2}$$

when the value of d is substituted in the third simultaneous equation. As the numerical values of a and b must remain unknown, the next best strategy is to eliminate one of them by expressing it in terms of the other. Since b is the same as $(-\frac{1}{2} - a)$ we can substitute this expression, which leaves the final equation with only one unknown quantity (a):

$$t = C\, m_s^a\, m_p^{(-\frac{1}{2} - a)}\, r^{3/2}\, G^{-\frac{1}{2}}$$

Now this complicated equation needs to be tidied up. This is done by seeing that:

$$m_s^a\, m_p^{(-\frac{1}{2} - a)} = m_s^a\, m_p^{-\frac{1}{2}}\, m_p^{-a} = \left(\frac{m_s}{m_p}\right)^a \frac{1}{\sqrt{m_p}}$$

and that:

$$r^{3/2}\, G^{-\frac{1}{2}} = \sqrt{\frac{r^3}{G}}$$

so the simplified version is:

$$t = C\sqrt{\frac{r^3}{Gm_p}}\left(\frac{m_s}{m_p}\right)^a$$

The cumbersome square root can be eliminated also in the interests of tidiness by squaring both sides:

$$t^2 = C^2\, \frac{r^3}{Gm_p}\left(\frac{m_s}{m_p}\right)^{2a}$$

This is only a partial solution because dimensional analysis can go no further in finding the numerical value of the exponent a. It is worth noting that the unknown term in the equation is a dimensionless number, being the ratio of two masses, so it has no units to affect the rest of the equation. The complete solution with the unknown term fully worked out can be found only by observations of the orbits of planets, or by deduction from other laws of physics. The partial solution is much better than no solution at all, however. It shows that the square of periodic time is a function of the cube of the major axis of the orbit; a rather unusual relationship which is, in fact, Kepler's law. Using dimensional analysis (which has involved only a minimal understanding of physics and no more than simple algebra), Kepler could have saved himself a considerable amount of trouble!

4.3 Drainage density

We return to Earth and the field of geomorphology for the third example of dimensional analysis giving a partial, but not a complete, solution to a problem. Drainage density (the total length of streams per area) is a variable which assumes very different values under different physical conditions. Where rainfall is very heavy and there is not much infiltration, a substantial amount of runoff occurs on the surface and many drainage channels result. In areas of weak rock or unconsolidated sediments many more channels develop than in areas of more resistant rock. Steep slopes are dissected by more channels than gentle slopes. But how can these commonplace observations be expressed in a theoretical equation?

The first stage is to define the variables that are thought to affect drainage density. The amount of rainfall running off the surface of the land can be found by subtracting the infiltration capacity of the land from the rainfall (both expressed in volume of water per area of land per time). This gives a measure known as the runoff intensity, whose dimensions are $[L^3L^{-2}T^{-1}]$ or $[LT^{-1}]$. The resistance of the surface to erosion is measured by the erosion proportionality factor, which gives the rate of material eroded per unit area divided by the eroding force applied per unit area. Dimensionally this is $[ML^{-2}T^{-1}/ML^{-1}T^{-2}]$ which simplifies to $[L^{-1}T]$. Finally, the steepness of slopes is most easily measured by relief, the difference in elevation between the highest and lowest points in the study area.

Setting out the dependent variable (drainage density) and the three independent variables that are thought to affect it:

D	drainage density	$[L^{-1}]$
Q	runoff intensity	$[LT^{-1}]$
k_e	erosion proportionality	$[L^{-1}T]$
H	relief	$[L]$

The power function relationship may be written:

$$D = C\ Q^a\ k_e^b\ H^c$$

$$[L^{-1}] = [1]\ [LT^{-1}]^a\ [L^{-1}T]^b\ [L]^c$$

Balancing the exponents of $[L]$: $-1 = a - b + c$

$[T]$: $0 = -a + b$

The simultaneous equations reveal that $\underline{a} = \underline{b}$ and $\underline{c} = -1$, but a complete solution for $\underline{a}$ and $\underline{b}$ is not possible since there are more unknowns than equations. The partial solution is, however, much better than no solution at all, because we now know that drainage density is inversely proportional to relief, and the variables $\underline{Q}$ and $\underline{k_e}$ both need to be raised to the same power:

$$D = C\frac{1}{H}\ (Qk_e)^a$$

The dimensional requirements of the relationship would be met if a were 1, -1, 2, 3 or indeed any number at all. To find out what the exponent actually is in nature requires an experiment.

4.4 Particle settling

To determine the grain size composition of a sample of sediment, the first step is to separate out the sand by sieving. The silt/clay component is then made up into a suspension in a liquid, and the amounts of material in the different grain size categories can be calculated from the weights of sediment still in suspension at successive time intervals. This technique relies on different sizes of grain in the sediment having predictable settling speeds. Particle settling velocity depends not only on the size of the grains but also on the viscosity of the liquid and the difference in density between the grains and the liquid. The acceleration due to gravity might also be expected to enter the relationship, since it is the falling of particles that is of interest. To discover the precise relationship, the variables are first set out:

w	particle settling velocity	$[LT^{-1}]$
r	grain diameter	$[L]$
μ	viscosity of liquid	$[ML^{-1}T^{-1}]$
ρ	sediment density minus liquid density	$[ML^{-3}]$
g	acceleration due to gravity	$[LT^{-2}]$

Expressing the particle settling speed as a power function of the other variables:

$$w = C\ r^a\ \mu^b\ \rho^c\ g^d$$

$$[LT^{-1}] = [1]\ [L]^a\ [ML^{-1}T^{-1}]^b\ [ML^{-3}]^c\ [LT^{-2}]^d$$

The simultaneous equations are:

For [L] $1 = a - b - 3c + d$ (i)

" [T] $-1 = -b - 2d$ (ii)

" [M] $0 = b + c$ (iii)

Notice that we have only three simultaneous equations but four unknown quantities to evaluate. It is therefore clear that only a partial solution can be achieved, by working out the values of three of the unknown exponents relative to that of the fourth.

Since b occurs in all three simultaneous equations, it is simplest to express a, c and d in terms of b. From equation (iii) we see that $c = -b$. Rearranging equation (ii):

$$2d = 1 - b$$

$$d = \tfrac{1}{2} - b/2$$

Replacing c and d in equation (i) with these expressions:

$$1 = a - b - 3\,(-b) + (\tfrac{1}{2} - b/2),$$

which reduces to:

$$a = \tfrac{1}{2} - 3b/2$$

The partial solutions are therefore $a = (\frac{1}{2} - 3b/2)$, $b = b$, $c = (-b)$ and $d = (\frac{1}{2} - b/2)$ and the equation expressing the relationship between the variables can be written:

$$w = C\, r^{(\frac{1}{2}-3b/2)}\, \mu^{b}\, \rho^{-b}\, g^{(\frac{1}{2}-b/2)}$$

Since $r^{(\frac{1}{2} - 3b/2)} = r^{\frac{1}{2}} .\, r^{(-3b/2)} = \sqrt{r}\left(\frac{1}{r}\right)^{3b/2}$

and the same applies to the exponent of g, the expression can be simplified to:

$$w = C\,\sqrt{rg}\left(\frac{\mu}{\rho}\right)^{b}\left(\frac{1}{r}\right)^{3b/2}\left(\frac{1}{g}\right)^{b/2} \qquad \text{(iv)}$$

This, however, is not very helpful! The second term on the right of the equation seems to indicate that the particle settling velocity depends on the square root of the grain diameter, but w also depends on grain diameter raised to the power (-3b/2). The overall relationship between the two variables remains completely obscure.

In some relationships it may be possible to liberate one or two variables from unknown exponents, while the remainder of the independent variables remain with their exponents unsolved. Partial solutions of this nature are much better than nothing. But in this example all the independent variables are left with unknown exponents as a result of the partial solution. This often happens when there are more variables in the equation than primary dimensions. Dimensional analysis does not reveal anything about the relationship between the variables in this example.

Having reached this stage, the best strategy is often to give up the attempt at finding the relationship by dimensional analysis. The researcher can resort to experiment or some other theoretical method for deriving the correct equation. However, this particular problem can be solved, by extending the concept of dimensions. What is required is at least one extra primary dimension, so the number of simultaneous equations is increased. How this can be done is explained in the next section, before returning to the example.

4.5 Vector dimensions

As was said earlier, there is nothing sacrosanct about the most commonly used system of mass, length and time dimensions. These (or any other dimensions) are not necessarily the most fundamental measures, even though most other measures can be derived from them. In some circumstances, it is useful to regard even these as composite measures, made up from other quantities. The concept of mass, for example, may be applied in one of two ways in physics, either to denote a quantity of matter or to denote an amount of inertia. Huntley (1967) has shown that physics equations which include the two meanings of mass must balance each independently. Two distinct dimensions, quantity of matter $[M_\mu]$ and inertia $[M_i]$ can therefore be defined to replace that of mass [M]. Using the new dimensions, the density of an object, formerly $[ML^{-3}]$, now becomes the quantity of matter per unit volume $[M_\mu L^{-3}]$. Similarly a force which uses the concept of inertia, is now $[M_i\ LT^{-2}]$. Huntley has shown how this division of the mass division into two independent components is useful for clarifying some relationships in physics and engineering. Since the distinction between quantity of matter and inertia is rarely (or never?) critical in an equation used in geography, we will not pursue the topic here.

What is worth a little elaboration is Huntley's second suggestion: that the dimension of length can be regarded as a composite of more fundamental dimensions. As distance has some claim to being the single variable of most interest to geographers, disaggregation of length into component measures might well have some applicability in geography. Huntley's argument is that length is a scalar measurement (a scalar is a quantity which has magnitude only) which can be replaced by three distinct vector measures, which comprise both magnitude and direction. Any length can be expressed as a combination of measurements in three perpendicular directions. The three directions can be thought of as height, length and breadth. All lengths measured in the vertical direction can be defined as having the dimension $[L_Z]$. The two directions at right angles in the horizontal plane can be defined as $[L_X]$ and $[L_y]$. Any vertical length can only be equal to another vertical length - not to a length measured in another direction. The implication of this is that if $[L_Z]$ appears on one side of an equation, it must also appear on the other side. Similarly, the dimensions $[L_X]$ and $[L_y]$ must also balance for an equation to be dimensionally homogeneous. While it is usually not necessary to make this fine distinction (the ordinary length dimension is sufficient for checking or deriving most equations) sometimes it can give a bit of extra information.

Using the three vector dimensions of length, an area on the Earth's surface is no longer $[L^2]$ but $[L_X L_y]$. Altitude has the dimension $[L_Z]$. A volume involves multiplying length, breadth and height, that is $[L_X L_y L_Z]$. Density becomes $[ML_X{}^{-1}L_y{}^{-1}L_Z{}^{-1}]$. The acceleration due to gravity (measured only in the vertical) is $[L_Z T^{-2}]$.

4.6 Stoke's law of settling

Huntley has demonstrated that the three vector dimensions of length can can be applied to completely solve the problem of particle settling described in section 4.4. Using vector dimensions, the variables are set out as follows:

w	particle settling velocity	$[L_Z T^{-1}]$
r	grain diameter	$[L_X{}^{\frac{1}{2}} L_y{}^{\frac{1}{2}}]$
μ	viscosity of liquid	$[ML_Z{}^{-1}\ T^{-1}]$
ρ	sediment density minus liquid density	$[ML_X{}^{-1}L_y{}^{-1}L_Z{}^{-1}]$
g	acceleration due to gravity	$[L_Z T^{-2}]$

These vector dimensions need some explanation. Remembering that $[L_z]$ is vertical, that is, the direction of a falling object, it is clear that $[L_z]$ is the length component in the settling velocity, the viscosity which acts in opposition to the falling grains and in the acceleration due to gravity. The density of the grains and the liquid is simply their mass divided by the three component lengths $[ML_x^{-1}L_y^{-1}L_z^{-1}]$. The only difficulty is with the grain diameter: in which of the three component directions is this quantity significant? Certainly not in the $[L_z]$ or vertical direction - the vertical elongation of the grain will not affect the speed at which it falls. What matters is the horizontal cross-section perpendicular to the direction of movement. Since $[L_x]$ and $[L_y]$ are equally applicable to the grain diameter, and since the grain diameter must have the dimensions of length (not length squared) then the two horizontal vectors must be allowed to contribute half each, giving $[L_x^{\frac{1}{2}}\ L_y^{\frac{1}{2}}]$.

The usual analysis can now be made, using the new vector dimensions:

$$w = C\ r^a\ \mu^b\ \rho^c\ g^d$$

$$[L_zT^{-1}] = [1]\ [L_x^{\frac{1}{2}}L_y^{\frac{1}{2}}]^a\ [ML_z^{-1}T^{-1}]^b\ [ML_x^{-1}L_y^{-1}L_z^{-1}]^c[L_zT^{-2}]^d$$

Balancing $[L_x]$: $0 = \frac{1}{2}a - c$

" $[L_y]$: $0 = \frac{1}{2}a - c$

" $[L_z]$: $1 = -b - c + d$

" $[T]$: $-1 = -b - 2d$

" $[M]$: $0 = b + c$

The five simultaneous equations are really four, because two of them are identical. Nevertheless, with four equations and four unknowns a clear solution is now possible:

$$a = 2,\ b = -1,\ c = 1,\ d = 1$$

So:

$$w = C\ \frac{r^2\rho g}{\mu}$$

The relationship is now explicit. You might be interested to substitute -1 for <u>b</u> in the partial solution given in section 4.4, to verify that that singularly uninformative equation is, in fact, quite correct. The final solution is Stokes' law, one of the fundamental relationships used in sedimentology. Although we have had to define new dimensions in order to wriggle out of seemingly impossible problems, this proves to be yet another example of the ability to derive laws by dimensional analysis.

5. SETS OF DIMENSIONLESS PRODUCTS

5.1 The π theorem

All the dimensionally homogeneous relationships we have examined so far can be rearranged so that both sides of the equation are dimensionless. For instance, the relationship between channel frequency F and drainage density D:

$$F = C\,D^2$$

could, without losing any information, be written as:

$$\frac{F}{D^2} = \text{constant}$$

which is dimensionless on both sides. The term F/D^2 is called a dimensionless product as it is a combination of dimensions which all cancel out, leaving the product without any units:

$$\frac{[L^{-2}]}{[L^{-1}]^2}$$

Notice that two physical quantities (F and D) derived from one primary dimension [L] have been combined to make one dimensionless product.

Another example is the relationship between air pressure (p), the height (h) of a column of mercury in a barometer, the density (ρ) of mercury and the acceleration due to gravity (g):

$$p = Cg\rho h$$

which is the same as

$$\frac{p}{g\rho h} = \text{constant}$$

Here, four quantities (p, g, ρ and h) derived from three primary dimensions ([M], [L] and [T]) have combined together to make a dimensionless product:

$$\frac{p}{g\rho h}$$

Let us now examine two relationships for which we could obtain only partial solutions using dimensional analysis. The periodic time (t) of the orbit of a planet was found to be related to the mass of the sun (m_S), the mass of the planet (m_p), the major axis of the orbit (r) and the universal gravitational constant (G). This relationship can also be expressed as an equation dimensionless on both sides:

$$\frac{r^3}{t^2Gm_p}\left(\frac{m_S}{m_p}\right)^{2a} = \text{constant}$$

We now have two dimensionless products, $\dfrac{r^3}{t^2Gm_p}$ and $\dfrac{m_S}{m_p}$.

These two products were formed by combining five quantities derived from three primary dimensions ([L], [M], and [T]).

The second relationship which was only partially derived was between drainage density (D), runoff intensity (Q), the erosion proportionality index (k_e) and relief (H):

$$D = C \frac{1}{H} (Qk_e)^a$$

Once again, the equation is made dimensionless by rearranging:

$$\frac{1}{HD} (Qk_e)^a = \text{constant}$$

There are two dimensional products in the relationship, $\frac{1}{HD}$ and (Qk_e). This time, the two dimensional products result from the combination of four physical quantities based on only two primary dimensions ([L] and [T]).

These examples lead to two observations; known as Buckingham's π theorem:

(1) A dimensionally homogeneous equation can be rewritten as a relationship between powers of dimensionless products.

(2) The number of dimensionless products is equal to the number of physical quantities minus the number of primary dimensions.

5.2 The π theorem and fluvial erosion

The advantage of the π theorem is that it makes it possible to simplify the interlinkages in a system of variables by reducing the variables to a manageable number. To follow through the last example, the original equations describe the characteristics of fluvially eroded landscapes according to the interrelationships of four variables: drainage density (D), runoff intensity (Q), erosion proportionality (k_e) and relief (H). But when four variables are interdependent it is not easy to disentangle what is happening. The relationship between two variables can be expressed in a graph, but three variables changing simultaneously can be represented only by a graph with three axes. Four variables working all at once are beyond the normal imagination. What Buckingham's π theorem offers is the possibility of reducing the number of variables, n, to a smaller number, (n-k), of new dimensionless variables, where k is the number of primary dimensions involved. In this example, it means reducing a function of four variables to a function of only two dimensional products. A function of two quantities can, of course, be completely appreciated from a single graph.

In mathematical terms, we have a functional relationship between the four quantities (bringing everything to the left hand side of the equation, so that nothing remains on the right):

$$f(D, Q, k_e, H) = 0$$

where f means ' a function of'. The π theorem says that this can be expressed as a functional relationship between just two variables, π_1 and π_2, which are dimensionless products of the original quantities.

$$f(\pi_1, \pi_2) = 0$$

Since Buckingham first expounded his theorem, it has been conventional to use the symbol π to represent any dimensionless product. In this context, π has no relation to its namesake, the constant ≈ 3.1416 in geometry.

The two dimensionless products in the fluvially eroded landscape example are:

$$\pi_1 = HD$$

and

$$\pi_2 = Qk_e$$

The first product, which is produced by multiplying relief by drainage density, has been called the ruggedness number by Strahler (1958). Landscapes with a high value for HD have a combination of relatively great relief and high drainage density. They consequently have steep slopes. Landscapes with low values for the index HD, on the other hand, are characterised by gentle slopes because relief is low and streams are few and far between. The second dimensionless product Strahler calls the Horton number. It indicates the intensity of erosion in the landscape. When the rainfall runoff intensity Q is high and the rate of removal of material k_e is also high, then π_2 is high and erosion is severe. When erosion is mild, π_2 is much lower. When examined together, the ruggedness number and the Horton number can give just as much information on the interplay of different factors in fluvial erosion as the four original variables. The π theorem does not tell us how the two numbers are related: to find that out, empirical data are required.

Strahler (1958) has taken this example further, by including all the variables which he considered to be causally related to drainage density. Strahler's list of variables was as follows:

D	drainage density	$[L^{-1}]$
Q	runoff intensity	$[LT^{-1}]$
k_e	erosion proportionality factor	$[L^{-1}T]$
H	relief	$[L]$
ρ	density of fluid	$[ML^{-3}]$
μ	viscosity of fluid	$[ML^{-1}T^{-1}]$
g	acceleration of gravity	$[LT^{-2}]$

According to the π theorem, the functional relationship

$$f(D,Q,k_e,H,\rho,\mu,g) = 0$$

can also be expressed as a relationship between four dimensional products:

$$f(\pi_1, \pi_2, \pi_3, \pi_4) = 0$$

There are four dimensional products now since there are seven physical quantities derived from three primary dimensions, and 7-3 = 4. The products may be identified as:

$\pi_1 = HD$ (ruggedness number)

$\pi_2 = Qk_e$ (Horton number)

$\pi_3 = \frac{HQ\rho}{\mu}$ (Reynolds number)

$\pi_4 = \frac{Q^2}{Hg}$ (Froude number)

The two additional dimensionless products which are required to incorporate the extra variables are both indices of fluid flow well known to hydrologists. The Reynolds number expresses the ratio of inertial force to viscous force. The Froude number gives the ratio of inertial force to gravitational force. You can check that all these numbers are dimensionless by writing them out in terms of the dimensions of their component variables and then cancelling.

5.3 Finding a set of products

How are dimensionless products recognised from a list of variables? The method is basically the same as the procedure for partially solving relationships.

Each dimensional product is the product of powers of the original quantities:

$$\pi = D^a,\ Q^b,\ k_e{}^c,\ H^d,\ \rho^e,\ \mu^f,\ g^k$$

Writing this in dimensions:

$$\pi = [L^{-1}]^a\ [LT^{-1}]^b\ [L^{-1}T]^c[L]^d[ML^{-3}]^e[ML^{-1}T^{-1}]^f[LT^{-2}]^h$$

In order for π to be dimensionless, the exponents of the primary dimensions must all cancel out, so:

$$\text{for } [L]:\quad 0 = -a + b - c + d - 3e - f + h$$

$$\text{for } [T]:\quad 0 = -b + c - f - 2h$$

$$\text{for } [M]:\quad 0 = e + f$$

These three simultaneous equations are impossible to solve fully since they contain seven unknowns. The way round this problem is to give arbitrary values to four of the exponents so that only three remain unknown. These can then be solved in relation to the arbitrarily fixed exponents. The result will be dimensional products which are internally consistent, but whose interrelationships are unknown and therefore open to empirical investigation.

To demonstrate, let us arbitrarily select the exponents, <u>a</u>, <u>b</u>, <u>d</u> and <u>h</u>. We could give them <u>any</u> values, but to keep simple we set a = 1, b = 0, d = 0 and h = 0. All these values can then be inserted in the appropriate places in the three simultaneous equations, which now simplify to:

$$[L]:\quad 0 = -1 - c - 3e - f$$

$$[T]:\quad 0 = c - f$$

$$[M]:\quad 0 = e + f$$

with only three unknowns now, the equations can be solved. The solutions are c = 1, e = -1 and f = 1. Remembering that we set a = 1 and the remaining exponents to zero, we can write out the first dimensional number:

$$\pi(i) = D^1,\ Q^0,\ k_e{}^1,\ H^0,\ \rho^{-1},\ \mu^1,\ g^0$$

Therefore

$$\pi(i) = \frac{Dk_e\mu}{\rho}$$

To check that this number is, indeed, dimensionless, write out the dimensions:

$$\frac{[L^{-1}]\ [L^{-1}T]\ [ML^{-1}T^{-1}]}{[ML^{-3}]}\ ,$$

which cancels completely.

By arbitrarily setting values for four of the exponents we have succeeded in deriving a combination of the variables that is dimensionless. Note, however, that it is not the same as any of the dimensionless products suggested by Strahler. The three other dimensionless products which go together with $\pi(i)$ are found by assigning different values to the selected exponents.

For the second product we set $\underline{a} = 0$, $\underline{b} = 1$, $\underline{d} = 0$ and $\underline{h} = 0$. The three simultaneous equations are now:

$$[L]:\quad 0 = 1 - c - 3e - f$$

$$[M]:\quad 0 = -1 + c - f$$

$$[T]:\quad 0 = e + f$$

with solutions: $\underline{c} = 1$, $\underline{e} = 0$ and $\underline{f} = 0$. Therefore,

$$\pi(ii) = Qk_e$$

The third product is given when $\underline{a} = 0$, $\underline{b} = 0$, $\underline{d} = 1$ and $\underline{h} = 0$. Making these substitutions gives the equations:

$$[L]:\quad 0 = -c + 1 -3e -f$$

$$[M]:\quad 0 = c - f$$

$$[T]:\quad 0 = e + f$$

which are solved by $\underline{c} = -1$, $\underline{e} = 1$ and $\underline{f} = -1$, to give:

$$\pi(iii) = \frac{H\rho}{k_e\mu}$$

Finally, to find the fourth product, set $\underline{a} = 0$, $\underline{b} = 0$, $\underline{d} = 0$ and $\underline{h} = 1$:

$$[L]:\quad 0 = -c -3e -f + 1$$

$$[M]:\quad 0 = c - f - 2$$

$$[T]:\quad 0 = e + f$$

The solutions, $\underline{c} = 3$, $\underline{e} = -1$ and $\underline{f} = 1$ give the product:

$$\pi(iv) = \frac{k_e^3\mu g}{\rho}$$

5.4 Complete sets of products

It is now clear that this set of dimensionless products $\pi_{(i)} - \pi_{(iv)}$ is different from the set $\pi_1 - \pi_4$ suggested by Strahler. There is, however, a close association between them. Every one of Strahler's products can be obtained by making different combinations of the latest set:

$$\pi_1 = \pi_{(i)}\ \pi_{(iii)}$$

$$\pi_2 = \pi_{(ii)}$$

$$\pi_3 = \pi_{(ii)}\ \pi_{(iii)}$$

$$\pi_4 = \pi_{(ii)}\ \pi_{(iii)}^{-1}\ \pi_{(iv)}^{-1}$$

While it is possible to derive any of the products from the products of the other set, no product can be obtained from the remaining three products in its own set. Each set of four dimensionless products is called a <u>complete set</u> if <u>each</u> product in the set is independent of the others and if <u>every other</u> dimensionless product of the variables can be obtained from <u>combinations of products</u> in the set. We could go on identifying new dimensionless products of these particular seven variables, but we would never find one that could not be produced by multiplying combinations of π_1 to π_4 or, alternatively, $\pi(i)$ to $\pi(iv)$. We have therefore identified two complete sets of dimensionless products. According to whichever set we choose to use, we can express the relationships between variables in a fluvially eroded landscape either with Strahler's set:

$$f\left(HD,\ Qk_e,\ \frac{HQ\rho}{\mu},\ \frac{Q^2}{Hg}\right) = 0$$

or with the derived set:

$$f\left(\frac{Dk_e\mu}{\rho},\ Qk_e,\ \frac{H\rho}{k_e\mu},\ \frac{k_e^3\mu g}{\rho}\right) = 0$$

What are still unknown are the exponents of the dimensional products when they are related to each other, so determining these becomes the task for empirical studies.

We could, of course, extract other complete sets of products simply by giving arbitrary values to different groups of four exponents in the original functional equation. With a large number of complete sets to choose from, how do we know which set is best? The answer is the set of products that can be most sensibly interpreted. Here, Strahler's set is the best as the first two products have unambiguous geomorphological significance, while the second two represent hydrological variables whose properties are already well known. The derived set, by comparison, is more difficult to interpret. The algebraic method for deriving sets of products in fact is a very inefficient way of arriving at a set which makes sense. Experienced analysts can usually pick out the most convenient dimensionless products simply by inspecting the list of variables. Provided (n - k) products are made up (where n is the number of variables and k is the number of primary dimensions) and provided that all the variables are included at least once, a set is complete.

5.5 Multivariate analysis and dimensionless products

The advantage of studying the relationships between dimensionless products rather than the relationships between the original variables is similar to the justification for using factor analysis or principal components analysis. In geography it is usually difficult to isolate a simple relationship between two variables by holding all other influences constant. More often, geographers are faced by systems of variables with complex interlinkages. Component or factor analysis is frequently used in these situations to identify clusters of variables that can be considered together and to replace a large number of variables by a smaller number of new composite quantities. Dimensional analysis can do the same job. Not only is the system of variables simplified by reducing the number of variable quantities to a few dimensionless products but the behaviour of the system may be studied independently of the units used to measure the original variables. The main advantage of using dimensionless products rather than factors is that the products are unambiguously interpretable and they are precisely defined so that exactly the same products can be investigated in quite different data sets. The disadvantage is that dimensional analysis requires a much greater prior understanding of the subjects of the research than factor analysis and its allied techniques. Whereas factor analysis is perhaps best used in an exploratory way, dimensional analysis has value only when all the variables which are relevant to the research can be specified with some confidence. Furthermore, variables used to create dimensionless products should contain a range of different secondary dimensions. A list of variables all measured in the same units would create a most uninteresting set of dimensionless products.

The work over several years by Wong on predicting mean annual flood in New England drainage basins is a good illustration of the superiority of a more theoretical dimensional analysis approach over the purely empirical principal components analysis method. Wong's first paper (1963) reported a study of 90 drainage basins in which the following variables were measured:

Q	mean annual flood	$[L^3T^{-1}]$
A	drainage area	$[L^2]$
I	precipitation frequency intensity	$[LT^{-1}]$
H	basin shape	$[L^3]$
M_t	mean altitude	$[L]$
S_D	stream density	$[L^{-1}]$
L_w	length of longest water course	$[L]$
L_s	length of main stream	$[L]$
S_c	main channel slope	$[1]$
T_c	tributary channel slope	$[1]$
S_A	average land slope	$[1]$
S_t	percentage area in ponds and lakes	$[1]$

First, Wong tried to relate mean annual flood to all the other variables using multiple regression, but he ran into the problem of multicollinearity.

That is, because the so-called independent variables were not independent but highly correlated with each other, the coefficients in the final regression equation were extremely untrustworthy outside the range of the data used. Wong therefore performed a standard principal components analysis on all the variables to sort them into clusters of similar variables. The largest cluster (component 1) was mostly of variables measuring size, such as drainage area and length of main stream, although basin shape was also a prominent member. Wong interpreted the second component as a 'slope' cluster, but as well as the slope variables it contained ingredients of mean altitude and precipitation frequency. Fortunately all other components were of minor importance. This, then, was a typical component or factor analysis result in which a slightly untidy grouping of variables could be roughly interpreted, with a bit of faith! Wong used the analysis to identify two variables which were unrelated to each other and which could stand as representatives for the two main clusters: the length of the main stream and the average land slope. Mean annual flood was related to these two variables using multiple regression to calculate the coefficients:

$$Q = k\, L_S^{1\cdot 29}\, S_A^{0\cdot 97}$$

where k was an uninterpreted numerical constant. The multiple correlation coefficient was 0.89.

In his first paper, Wong made no attempt to achieve dimensional homogeneity in his equation. He had succeeded in finding an empirical equation which predicted annual flood size reasonably well, rather than a theoretical description that would help to explain why floods vary. But in 1978, Wong produced a second paper demonstrating that the twelve interrelated variables could be reduced to a set of ten dimensionless products:

$$\pi_1 = Q/AI \qquad \pi_2 = H/(\sqrt{A})^3$$
$$\pi_3 = M_t/\sqrt{A} \qquad \pi_4 = S_D/\sqrt{A}$$
$$\pi_5 = L_W/\sqrt{A} \qquad \pi_6 = L_S/\sqrt{A}$$
$$\pi_7 = S_C \qquad \pi_8 = T_C$$
$$\pi_9 = S_A \qquad \pi_{10} = S_t$$

In a third report, Wong (1979) worked out the numerical values of each of the dimensionless products for each of the New England drainage basins. Treating the products as ten new variables, he examined the strength of relationships between them and found that π_1 and π_9 were especially closely related. Expressing π_1 as a function of π_9, with the usual dimensional constant C:

$$\pi_1 = C\, (\pi_9)^a$$

or

$$\frac{Q}{AI} = C\, (S_A)^a$$

This meant that Q, the mean annual flood, could be expressed as a function of A, I and S_A in a dimensionally homogeneous equation:

$$Q = C(AI)\, (S_A)^a$$

A multiple regression using (AI) and S_A as the independent variables confirmed that the mean annual flood is related to AI (raised to the power 0.9, which Wong considered close to the theoretically expected 1) and S_A raised to the power of 1.

There are two reasons why the final equation is superior to the one Wong settled for in 1963. Firstly, it is a dimensionally balanced equation derived from physical reasoning, whereas the first analysis was not based on theoretical foundations. Secondly, while the original equation was fairly accurate in predicting $\bar{Q}$ (R = 0.89), the final version was even more accurate (multiple correlation coefficient R = 0.92). In other words, dimensional analysis helped Wong to find a relationship that was theoretically respectable and which fitted the data better than the original derived from principal components analysis. Not every multivariate study in geography could adopt this approach, of course (usually the variables are not varied enough), but where it is possible there are substantial advantages to be gained.

5.6 Dimensionless products in agricultural location

To work with dimensionless products, the investigator must be able to identify all the variables which interact with each other and, ideally, the variables should be made up of a variety of primary dimensions. These conditions are more difficult to achieve in human geography than in physical geography. This is not to say, though, that illustrations cannot be found in human geography. Take, for example, the agricultural location theory originating from von Thünen. Following from this theory, the connections between the value of agricultural land, the location of the land and productivity may be expressed as a relationship between the following variables:

E	yield per unit area	$[ML^{-2}T^{-1}]$
D	distance from market	$[L]$
R	rent per unit area	$[\$L^{-2}T^{-1}]$
q	gross profit per unit crop	$[\$M^{-1}]$
w	transport rate per unit crop	$[\$L^{-1}M^{-1}]$

By inspecting this list of variables, a number of dimensionless products can be made up:

$$\pi_1 = \frac{Eq}{R}, \qquad \pi_2 = \frac{EwD}{R}, \qquad \pi_3 = \frac{Dw}{q}$$

These three do not exhaust the possibilities, for each of them gives rise to many variations. The first product, for example, can be used to derive:

$$\pi_4 = \frac{R}{Eq}, \qquad \pi_5 = \left(\frac{Eq}{R}\right)^2 \qquad \pi_6 = \left(\frac{Eq}{R}\right)^3$$

and so on, none of which differs substantially from π_1, which measures the gross profit as a proportion of land rent for a unit area. The second product, π_2, expresses transport costs as a proportion of land rent, while π_3 measures the transport costs as a fraction of the gross profit of the crop. The three products are interdependent, since $\pi_1 = \frac{\pi_2}{\pi_3}$

A combination of any two of them can be used to derive the third, so any two of them make up a complete set of products. This is as expected, as there are five variables and three primary dimensions, so the number of products in a complete set is 5-3 = 2. It is up to the researcher to choose whichever pair of products is thought to be the most interesting. Empirical data can then be used to produce a graph of one dimensionless product against the other, which will portray the essence of the interlinkages between five variables. By graphing π_1 against π_2 for instance, it would be possible to study the relationship between gross profits and transport costs for different crops, while holding land area and rent constant.

6. PROBLEMS OF SCALE

Geographers are accustomed to studying similar phenomena at different scales. Human settlements, for example, might consist of a few buildings around a route intersection or cities of millions of buildings spread over very large areas of the Earth's surface. The streams of interest to geomorphologists vary in size from insignificant first order creeks too small to be mapped to great river systems covering sub-continents. Other natural systems or engineering works, perhaps extending over many square kilometers in reality, are subjected to experiments in the form of models small enough for the laboratory bench. With such enormous variations as these, it becomes likely that the properties of the objects under study do not remain the same, but change with differences in size. Yet another application of dimensional analysis is in understanding the implications of such changes.

6.1 Geometrical similarity

If two objects differ in size so that one has linear measurements which are all a constant proportion of the corresponding linear measurements of the other, the objects are said to be geometrically similar. The larger object is simply a magnified replica of the smaller. Geometrical similarity might apply to a single object which changes size through time, or it might also apply to several objects in different sizes at the same time.

Objects which are geometrically similar are the same shape. If two geometrically similar objects have linear dimensions in the proportions 2:1, then corresponding measures of area will be in the ratio 4:1. Their volumes will be in the proportions 8:1. That is, their linear, areal and volumetric measurements will be proportional to the dimensions $[L]$, $[L^2]$ and $[L^3]$ respectively.

So long as the condition of geometric similarity is met, it is possible to predict the relationships between the properties of size extremely easily, just by referring to the dimensions of the properties. For example, it should come as no surprise that the length of the longest stream in a drainage basin is related to the area of the basin. If drainage basins are geometrically similar (if large basins are the same shape as small basins) then the relationship between the lengths (ℓ) and areas (A) of several drainage basins must be:

$$\ell = C\,A^{0.5}$$

where C is a dimensionless scaling constant. Length must be proportional to the square root of area to preserve the dimensional balance of the equation. If area is related to the length of perimeter of the basin (p), the expected equation is:

$$A = C\ p^2$$

for the same reason. Here C represents another dimensionless constant of proportionality. Measures of length are directly related to each other in conditions of geometrical similarity, so for instance the wavelength of stream meanders (λ) can be predicted to vary in proportion to the width of the stream channel (z) according to the simple relation:

$$\lambda = C\ z$$

In the same way, areal measures are expected to be directly related to other areal measures: the area of alluvial fan deposited by a river is directly proportional to the catchment area of the river, provided the conditions of geometrical similarity are met. The depth of the alluvial deposits, however, will be related to the square root of the drainage area, and the volume of alluvium (v) will vary with the drainage area (A) raised to the power of 1.5 in order to achieve dimensional homogeneity:

$$v = C\ A^{1.5}$$

$$[L]^3 = [1][L^2]^{1.5}$$

6.2 Allometric relationships

Geometrical similarity is, in fact, comparatively rare in nature. It is physically impossible for organisms and even many man-made objects to grow uniformly in size and preserve geometrical similarity while still functioning in the same way. As an organism grows, some parts of it increase in size more rapidly than others. In other words, it changes its shape (Thompson, 1961; Gould, 1966).

Allometric relationships are the differences in proportions associated with changes in absolute magnitude. Huxley (1932) proposed a simple equation to describe these relationships, which has become known as the equation of simple allometry:

$$y = bx^{\alpha}$$

where y is the magnitude of part of the organism and x is the magnitude of another part of the organism or, more commonly, a measure of total organism size. The symbols b and α represent constants, with α being the ratio of the growth rates of x and y. This equation is a useful link between the ideas of relative growth in biology and relationships associated with size in geography. While it was intended to describe the growth of a single organism through time, it can equally well be applied to several differently-sized objects at the same time.

For example, the relationship between the length of the longest stream (ℓ) and the drainage area (A) can be written:

$$\ell = b\ A^{\alpha}$$

We have already deduced that the value of the exponent α is expected to be 0.5 under conditions of geometric similarity. Geometric similarity is therefore a special case of the allometric relationship, which occurs in this example when $\alpha = 0.5$. It is likely, however, that researchers who actually measured the lengths and drainage areas of several streams and subjected the results to regression analysis would find that α is rarely exactly 0.5. A typical real relationship might be:

$$\ell = 1.4A^{0.64}$$

It is clear that these particular drainage basins are not geometrically similar. Basins of a different size are also a different shape. Firstly, large drainage basins tend to be more elongate than small drainage basins and, secondly, the sinuosity of the main stream (and also the accuracy of its cartographic representation) changes with increasing scale.

This demonstrates how variations from the exponents which are expected using simple dimensional arguments can indicate changes in shape associated with size differences. To use another river basin example, the area of the basin (A) can be related to its perimeter length (p) in the equation:

$$A = C\,p^{\alpha}$$

If there is no shape change, we would expect the value of α to be 2. If an empirical analysis gave a different exponent, say $\alpha = 1.7$, this would reveal an allometric relationship: the shape of the basins changes with size. In this case, larger stream systems would have more crenulated basin peripheries (hence relatively longer perimeters) relative to their size than small stream systems. Church and Mark (1980) extend this type of analysis.

The study of relationships associated with the size of cities suffers from the same sort of problem. If large cities were simply scaled-up versions of small cities we would be able to predict the relationship:

$$a = \frac{p}{w}$$

where a is the built up area of the city, p is the population of the city and w is the population density. This follows from the dimensions of the variables:

$$[L^2] = \frac{[N]}{[NL^{-2}]}$$

Area is directly proportional to population so the exponent of population is one. However the real situation is not as simple as this. It is much more likely that city area will vary with population raised to a power less than one. This is because large cities are not simply scaled-up versions of small cities. The most notable difference in this context is that average population density increases with city size, so large cities are more compact than small cities. The simple dimensional equation is apt to be misleading if it is taken at face-value. But, as with the river basin examples, once it is realised that geographical systems of different sizes are not necessarily geometrically similar, dimensional arguments can be used to spotlight the effects of scale on shape and form.

6.3 Scale models

Dimensional analysis is also helpful in examining the effects of scale in experimental models. It is often very difficult to perform experiments on natural processes which occur over a large area of the Earth's surface or over a long time period. In these circumstances it may be more convenient to make a working model and do the experiments on that. The model may be a scaled down version of the real world, and sometimes it is speeded up as well. Models have been made of atmospheric and oceanic circulation systems, hill slopes, river channels, tidal estuaries and glaciers. Having built a model which behaves similarly to the natural process, the prevailing conditions can be altered by making adjustments to the model and the effects can be observed. Transferring the results of the model test to the process operating in the real world is not necessarily straightforward, because the change in scale from the real world to the model can distort the processes under observation. Dimensional analysis is used to reduce such distortion to a minimum.

The simplest kind of scale model is one in which everything in the model is a constant fraction of the size of the corresponding real feature. Complete geometrical similarity is not always easy to achieve because if the model contains sediment, for example, the grain size is likely to be much larger in proportion to the overall linear dimensions of the scaled down model than in the real world. This may lead to serious problems of surface roughness and friction in the model which are negligible in the natural system. Geometrical similarity may actually be undesirable in cases where one linear dimension needs to be exaggerated to emphasise a particular characteristic. Scale models of coastlines or estuaries, for example, usually have the vertical length dimension exaggerated, so that tides and waves can be simulated more effectively. In that case the dimensions $[L_x]$ and $[L_y]$ would be in a certain ratio between model and reality, while the dimension $[L_z]$ would be designed to have a different ratio.

Just as it is convenient to make the model a smaller version of reality so that it is more accessible for experimentation, it is often desirable to make the model faster in operation than reality, so that processes occupying days or even many years can be observed in a much shorter time period. So long as the time dimension in the model is a constant fraction of the time dimension in reality, results can be extrapolated from one to the other. Models in which lengths and times are both held in constant proportions with reality are kinematically similar. Full dynamic similarity is achieved when the dimensions of length, time and mass are all strictly proportional between model and reality. Forces operating in the model then bear a constant relationship to forces in reality, a relationship which can be calculated from the ratios between model and reality in the mass, length and time dimensions.

If the model is to be a smaller, faster version of reality, some of the properties of the materials used may need to be altered. To simulate a very small glacier moving much faster than a real glacier, for instance, it is necessary to use a material much less viscous than ice. The substitute material must, however, be rigid enough to crack or form crevasses when subject to tensile stresses, while yielding under shear stress. A mixture of kaolin and water has been used for this purpose.

When the mass, length and time dimensions are not equal to those in the real world (although they may each be proportional) interpreting results from the model must be done with care. Much the simplest way to do this is to eliminate the effects of dimensions and units by working with dimensionless products. The properties of the model must be adjusted so that all the relevant dimensionless products have the same value in both the model and the real world. For example, in a model of a drainage basin constructed to reproduce the relationships between drainage density, relief, runoff rates and erosion intensity found in the real world, the relevant dimensionless products would be those identified in an earlier section:

$\pi_1 = HD$ (ruggedness of landscape)

$\pi_2 = Qk_e$ (intensity of erosion)

$\pi_3 = \frac{HQ\rho}{\mu}$ (inertial force proportional to viscous force)

$\pi_4 = \frac{Q^2}{Hg}$ (inertial force proportional to gravitational force)

To keep the same values of the dimensionless products between model and reality means that the materials used must have certain properties. If the length dimension is to be reduced and time is to be speeded up, then both the runoff intensity (Q) and the erosion proportionality factor (k_e) will be affected. To preserve the balance between measures in the third dimensionless product, the density (ρ) and viscosity (μ) of the fluid may need to be different between model and reality, so it may not be desirable to use water for the 'rainfall' in the model. After making these adjustments to ensure that the first three dimensionless numbers are identical between reality and model it may be impossible to do the same for the fourth, which contains gravity (a value not easily manipulated). The investigator would therefore not be able to extrapolate the fourth dimensionless measure directly from model to real world, although the other three measures would be immediately interpretable.

Using dimensional analysis, the art of scale model design is in selecting dimensionless numbers which capture the aspect under investigation and in matching these numbers in model and reality. Dimensional analysis helps in this way to keep the most serious effects of scale changes within experimental control.

7. CONCLUSION

Dimensional analysis is applicable in geography whenever mathematical equations are used to express the relationships between quantities. Algebraic mistakes or errors of omission can be checked using the method, forgotten equations can be reconstructed and ambiguously defined variables can be clarified. Changes from one set of units to another are facilitated. When variables are related together in a power function, the dimensional method will often reveal the numerical value of all the exponents in the equation. When this is not possible, a useful partial solution can usually be produced. Networks of complex interdependencies within systems of variables can be untangled by isolating dimensionless products, and dimensional analysis is used to show where empirical investigations are necessary in order to discover the values of constants. The method is also indispensable when considering the distorting effects of scale and in interpreting the results of experiments with scale models.

The basic dimensional method is not difficult to learn, but neither is it always easy to apply in geography. Perhaps the most serious difficulty is that accurate identification of the relevant variables is a necessary precondition for dimensional analysis. Some branches of geography have not yet developed a coherent body of theory which allows the investigator to specify the variables that are relevant to a particular problem. This is especially so in human geography, where the definition and measurement of behavioural and psychological quantities is not well established. Another drawback is that the method is most effective when applied to power function relationships. While many equations in geography are of this general type (as, for example, when a straight line relationship is fitted to two variables measured on logarithmic scales), a common exception is the exponential form of association. Dimensional analysis is useful in checking, but not in deriving, exponential relationships.

Perhaps the best justification for dimensional analysis is that it traces back the processes we observe in the world to our definitions and methods of measurement. The mathematical equations we thought were the result of painstaking fieldwork and statistical analysis turn out to be entirely predictable from the definitions and measurements employed. Equations become less mysterious and much more interpretable in this light. Dimensional analysis is not just another quantitative abstraction thrust upon geography. It is a way of bringing the mathematics down to earth.

8. BIBLIOGRAPHY

Many books give more advanced descriptions of dimensional analysis than has been possible here but they do not, unfortunately, contain many examples of direct interest to the geographer. While there are exceptions (notably in biology and economics) most of them deal with physics and engineering problems. Out of those available, I personally have found Huntley and Douglas most helpful. I have included these with a selection of other general texts, together with a few relevant research papers.

Allen, J. (1947). *Scale models in hydraulic engineering*, Longmans, London.

Bridgeman, P.W. (1922). *Dimensional analysis*, Yale University Press, New Haven.

Buckingham, E. (1914). On physically similar systems: illustrations of the use of dimensional equations. *Physical Review*, 4, 345-376.

Church, M. & Mark, D.M. (1980). On size and scale in geomorphology. *Progress in Physical Geography*, 4, 342-390.

Corrsin, S. (1951). A simple geometrical proof of Buckingham's π-theorem. *American Journal of Physics*, 19, 180-181.

Douglas, J.F. (1969). *An introduction to dimensional analysis for engineers*, Pitman, London.

Duncan, W.J. (1953). *Physical similarity and dimensional analysis: an elementary treatise*, Edward Arnold, London.

Gould, S.J. (1966). Allometry and size in ontogeny and phylogeny. *Biological Review*, 41, 587-640.

Haynes, R.M. (1975). Dimensional analysis: some applications in human geography. *Geographical Analysis*, 7, 51-67.

Haynes, R.M. (1978). A note on dimensions and relationships in human geography. *Geographical Analysis*, 10, 288-292.

Huntley, H.E. (1967). *Dimensional Analysis*, Dover Publications, New York.

Huxley, J.S. (1932). *Problems of relative growth*, Methuen, London.

Isaacson, E. de St Q. & M. de St Q. Isaacson. (1975). *Dimensional methods in engineering and physics*, Edward Arnold, London.

Jong, F.J. de (1967). *Dimensional analysis for economists*, North Holland Publishing Co., Amsterdam.

Jupp, E.W. (1962). *An introduction to dimensional method*, Cleaver-Hume Press, London.

Kline, S.J. (1965). *Similitude and approximation theory*, McGraw-Hill, New York.

Langhaar, H.L. (1951). *Dimensional analysis and theory of models*, John Wiley, New York.

Melton, M.A. (1958). Geometric properties of mature drainage systems and their representation in an E_4 phase space. *Journal of Geology*, 66, 35-56.

Palacios, J. (1964). *Dimensional analysis*, Trans.P.Lee, Macmillan, London.

Pankhurst, R.C. (1964). *Dimensional analysis and scale factors*, Chapman and Hall, London.

Stahl, W.R. (1961). Dimensional analysis in mathematical biology I: general discussion. *Bulletin of Mathematical Biophysics*, 23, 355-376.

Stahl, W.R. (1962). Dimensional analysis in mathematical biology II. *Bulletin of Mathematical Biophysics*, 24, 81-108.

Strahler, A.N. (1958). Dimensional analysis applied to fluvially eroded landforms. *Bulletin of the Geological Society of America*, 69, 279-300.

Thompson, D'Arcy W. (1966). *On growth and form* (abridged edition), Cambridge University Press, Cambridge.

Wong, S.T. (1963). A multivariate statistical model for predicting mean annual flood in New England. *Annals of the Association of American Geographers*, 53, 293-311.

Wong, S.T. (1978). Toward achieving dimensional homogeneity in hydrologic analysis. *Geographical Analysis*, 10, 262-272.

Wong, S.T. (1979). A dimensionally homogeneous and statistically optimal model for predicting mean annual flood. *Journal of Hydrology*, 42, 269-279.